Stereochemical Applications of NMR Studies in Rigid Bicyclic Systems

Methods in Stereochemical Analysis

Volume 1

Series Editor
Alan P. Marchand
**North Texas State University
Department of Chemistry
Denton, Texas 76203**

Stereochemical Applications of NMR Studies in Rigid Bicyclic Systems

By

Alan P. Marchand

Verlag Chemie International

Deerfield Beach, Florida

Alan P. Marchand, Ph.D.
Department of Chemistry
North Texas State University
Denton, Texas 76203

Library of Congress Cataloging in Publication Data

Marchand, Alan P.
   Stereochemical applications of NMR studies in rigid bicyclic systems.

   (Methods in stereochemical analysis; v. 1)
   Includes bibliographical references and index.
   1. Cyclic compounds—Analysis. 2. Nuclear magnetic resonance spectroscopy.
3. Stereochemistry. I. Title. II. Series.
QD400.M37 1982   547′.5046   82-13648
ISBN 0-89573-112-6

Printed in the United States of America.

ISBN 0-89573-112-6 Verlag Chemie International, Deerfield Beach
ISBN 3-527-25976-7 Verlag Chemie GmbH, Weinheim

The White Rabbit put on his spectacles. "Where shall I begin, please your Majesty?"

"Begin at the beginning," The King said, very gravely, "and go on 'till you come to the end: then stop."

—Charles Lutwidge Dodgson

SERIES FOREWORD

Methods in Stereochemical Analysis is intended to provide a format for critical and timely reviews that deal with the applications of physical methods for determining conformation, configuration, and stereochemistry. The term "stereochemical analysis" is interpreted in its broadest sense, encompassing organic, inorganic, and organometallic compounds as well as molecules of biochemical and biological significance. The methods include, but are not restricted to, *spectroscopic techniques* (e.g., NMR, infrared, UV-visible, Raman, mass, and optical spectroscopy), *physical techniques* (e.g., calorimetry, photochemical, kinetic, and "direct" methods such as X-ray crystallography, neutron and electron diffraction), and *applied theoretical approaches* to stereochemical analysis.

In establishing the series, the editor and members of the advisory board seek to attract contributions of the highest scientific caliber from outstanding investigators who are actively pursuing research on stereochemical applications of these various techniques and/or applied theoretical approaches. We envision that contributions might adopt either the form of a monograph or follow a multiauthor treatise format with individual chapters contributed by a number of outstanding research scientists. But whichever format is decided upon for a particular volume, we consider most desirable the inclusion of critical and timely reviews which place the author's own work in perspective with regard to other important literature in the field while at the same time retaining the highly personal character of his/her individual contributions. Special emphasis is placed upon the concept of *critical* and *timely* (rather than necessarily comprehensive) reviews in this regard.

Whatever merit the resulting volumes possess necessarily must derive from the excellence of the individual contributions. Accordingly, the editor welcomes suggestions from members of the scientific community of potential topics for inclusion in the series and names of potential contributors, as well as suggestions of a critical nature which might assist him in better fulfilling the stated objectives. It seems fitting, therefore, that the series be dedicated to its readership among members of the scientific community, for ultimately *they* will gauge the degree to which the series fulfills its objectives.

Alan P. Marchand
Denton, Texas

PREFACE

Rigid bicyclic systems, particularly bicyclo[2.2.1]heptanes (norbornanes) and bicyclo[2.2.2]octanes, have occupied a special niche in the hierarchy of organic compounds. They are readily accessible synthetically through single step cycloaddition (Diels-Alder) reactions of cyclic dienes with appropriately substituted acyclic dienophiles. Such cycloaddition reactions often occur with pronounced endo stereoselectivity and always proceed with cis stereospecificity with respect to the orientation of substituents on the diene and the dienophile. In addition, a number of substituted norbornanes and bicyclo[2.2.2]octanes occur in nature. These facts account in part for the overwhelming popularity which these systems have enjoyed as substrates for mechanistic studies and as versatile synthetic intermediates.

With the advent of high-resolution NMR spectroscopy, it soon became apparent that specifically substituted norbornanes and bicyclo[2.2.2]octanes could fulfill an additional role. Their relatively rigid carbon atom frameworks and consequently their relatively fixed molecular geometries render them particularly useful as model systems for conformational and stereochemical analysis. Thus, substituted norbornanes and bicyclo[2.2.2]octanes have been utilized extensively as substrates in NMR investigations to elucidate phenomena as diverse as nonbonded substituent steric interactions, the mechanism of transmission of electronic substituent effects, and the geometrical dependence of spin–spin coupling between distant nuclei.

A wealth of information relating to NMR spectral properties of rigid bicyclic systems has accrued during the past 25 years. Indeed, the vast accumulation of literature dealing with various aspects of NMR studies on norbornanes and on bicyclo[2.2.2]octanes alone essentially precludes an exhaustive treatment of this subject in a review of any reasonable length. Instead, the present monograph seeks to highlight critically some of the more significant stereochemical applications of NMR chemical shifts, coupling constants, and related phenomena which have been reported in recent years for the norbornyl and bicyclo[2.2.2]octyl systems. Such self-imposed restrictions undoubtedly will result in the omission of some important individual contributions. Such omissions, when they occur, certainly do not reflect negative judgments on the part of the author.

The task of preparing the manuscript was aided materially by advice and comments freely offered by several friends and colleagues. In particular, the following investigators kindly provided manuscript preprints in advance of publication describing recent results of NMR studies performed in their own laboratories: William Adcock, Michael Barfield, George A. Olah, John L. Wong, Leslie D. Field, James L. Marshall, and G. R. Surya Prakash. Helpful and stimulating discussions with Michael Barfield and James L. Marshall are acknowledged gratefully. In particular, the author expresses his gratitude to Dr. Marshall for having read the entire manuscript and for having provided numerous helpful, critical comments.

Portions of the manuscript were written while the author was on sabbatical leave from The University of Oklahoma. Special thanks are extended to the Departments of Chemistry at the California Institute of Technology (Spring Semester, 1980) and at The University of Southern California (Summer, 1980), and also to the Chemistry Division, Naval Weapons Center, China Lake, California (Summer, 1981), all of whom extended their library facilities and their warm hospitality to the author.

Finally, I gratefully acknowledge the professional and personal contributions of my wife, Nancy Wu Marchand, who provided the tranquil atmosphere and, at times, the moral support which rendered a potentially onerous task both enjoyable and rewarding.

CONTENTS

1

INTRODUCTION

During the past 40 years, bicyclo[2.2.1]heptane (norbornane) and bicyclo-[2.2.2]octane derivatives have been exploited extensively by physical-organic chemists as model systems for elucidating chemical phenomena as diverse as the mechanism of Wagner-Meerwein (carbonium ion) rearrangements[1-3] and the mechanism of transmission of electronic substituent effects.[4,8] A primary virtue of these systems is the relative rigidities of their respective carbon atom frameworks, with the result that their molecular geometries can be estimated with reasonable accuracy. This in turn permits the relatively accurate estimation of stereochemical effects in these systems. Furthermore, potential contributions from conformations in complex acyclic systems can be approximated by appropriate patterns of functional group substitution in these rigid ring systems. The relative importance of the contribution of a particular conformation in the acyclic system then can be estimated via appropriate physical measurements on its rigid system analog.

Often, considerable synthetic skills are required to prepare isomerically pure, substituted bicyclo[2.2.1]heptanes and bicyclo[2.2.2]octanes. Attendant with their synthesis is the often difficult problem of structural characterization, especially in cases where the synthetic sequence employed may afford more than one isomeric product, e.g., exo and endo isomers. The difficulties encountered by early investigators in this regard could be surmounted more readily once proton nuclear magnetic resonance (NMR) instrumentation became available on a routine basis in the 1950s.

As investigators gained more experience, it rapidly became evident that the proton chemical shift and coupling constant correlations that were proving useful for structural elucidation in these rigid bicyclic systems might have more universal applicability. The decade of the 1960s witnessed an explosion

1

of interest in NMR studies on substituted norbornanes and bicyclo[2.2.2]-octanes. Stereochemical effects on proton chemical shifts and coupling constants determined in these rigid systems have indeed proved to be useful for elucidation of structure in much more complex macromolecules, many of which are of intense interest to biochemists and to molecular biologists who study life processes. The routine availability of ^{13}C NMR instrumentation in the 1970s has permitted investigators to gain information that pertains directly to the conformation of the carbon skeleton itself in these complex biomolecules. Once again, preliminary ^{13}C NMR studies in the norbornyl and bicyclo-[2.2.2]octyl systems have provided the basis for understanding related phenomena in the much more complex biomolecules.

Over the years, the norbornyl and bicyclo[2.2.2]octyl systems repeatedly have afforded the key model compounds on which are predicated many of the pioneering NMR studies utilizing rigid systems of known molecular geometry. Their importance in this regard cannot be overstated; these systems have truly held a unique position of importance to NMR spectroscopists and to practicing physical-organic chemists. It therefore would seem appropriate to review what has been learned from NMR studies of norbornanes, bicyclo[2.2.2]-octanes, and related systems, not only to place past work in perspective but it is hoped, to uncover suitable areas for future exploitation by NMR investigators.

2

AIDS TO MAKING NMR SPECTRAL ASSIGNMENTS IN RIGID BICYCLIC SYSTEMS

Lanthanide-Induced Shifts

Early proton NMR investigations of substituted norbornanes and bicyclo-[2.2.2]octanes, generally performed at 40 MHz, were hampered frequently by both spectral complexity and incomplete signal resolution. This adverse situation was ameliorated to a great extent with the advent of more sophisticated, high-resolution NMR instrumentation capable of operating at significantly higher magnetic field strengths and radiofrequencies (e.g., 100, 220, and 300 MHz, and beyond). Nevertheless, proton NMR spectra of substituted norbornanes in bicyclo[2.2.2]octanes are often sufficiently complex, even when observed at higher field strengths, that supplementary aids to spectral resolution and simplification have been sought.

One method that has been utilized successfully take advantage of the large magnetic anisotropy effects associated with aromatic molecules. When used as NMR solvents, benzenoid aromatic systems interact with solute molecules, producing aromatic solvent-induced shifts (ASIS) whose sign and magnitude for a particular proton or group of protons in the solute molecule are determined by their geometry relative to polar functional groups in that molecule which interacts strongly with the aromatic solvent.[6] Although the ASIS phenomenon was explained at first in terms of a model that postulated the existence of a discrete 1 : 1 solute–solvent "collision complex," an alternative explanation that invokes a general (time-averaged) solvation model appears to be gaining acceptance.[7,8]

Although ASIS in several instances has proved valuable in facilitating NMR spectral interpretation, the magnitudes of the shifts produced by this method are often too small to permit complete resolution of the kinds of complex proton NMR spectra that are generally encountered for substituted norbornyl and bicyclo[2.2.2]octyl systems. In cases where these compounds bear strongly (Lewis) basic functional groups, the method of choice for resolution of their NMR spectra has been to employ lanthanide shift reagents (LSRs).

The practical application of paramagnetic lanthanide β-diketoenolate complexes for inducing NMR spectral shifts was first demonstrated by Hinckley in 1969.[9] In succeeding years, a wide range of LSRs has been introduced; lanthanide acetylacetonate derivatives appear to be the most generally useful of these. The magnitude of pseudocontact (dipolar) shifts produced by added LSR depends in part on geometric factors. Hence, analysis of the lanthanide-induced shifts (LISs) thereby produced permits detailed stereochemical studies to be performed on the substrate molecule (solute), in addition to NMR spectral simplification and clarification. The underlying theory and practical applications of LSRs in NMR spectroscopy have been reviewed recently.[10-18]

Rigid bicyclic systems have proved popular as substrates in LIS studies, primarily because of the complexity of their normal (unshifted) proton NMR spectra and the corresponding need for additional spectral resolution and simplification.[17] Additionally, the rigidity associated with their respective carbon frameworks reduces the degree of uncertainty attendant with analyzing distance and geometric factors, thereby simplifying the task of assigning stereochemistry in these systems by NMR spectroscopy.

Figures 1 and 2 demonstrate the degree of NMR spectral resolution that can be attained via incremental addition of LSR to the NMR sample. Figure 1 presents the normal 60-MHz NMR spectrum of dimer ketone 1, which is synthesized by iron carbonyl-promoted coupling of 7-phenylnorbornadiene to carbon monoxide. The dimer ketone thereby obtained has been shown to possess the anti-exo-trans-exo-anti configuration (1).[19,20]

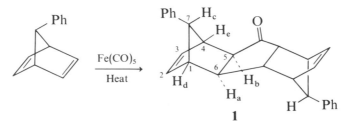

The incomplete resolution of H_c and H_d in the NMR spectrum shown in Figure 1 precludes assignment of the stereochemistry at C-7 in 1. Satisfactory resolution of all proton signals in 1 is achieved by the addition of Eu(fod)$_3$* (Figure 2, $\rho = $ [LSR]/[substrate] $= 0.250$). Figure 3 shows the effect of the [LSR]/[substrate] concentration ratio (ρ) on the magnitudes of the individual proton shift gradients ($\Delta\delta$ for H_a–H_e). Excellent linear correlations are found between these quantities for each proton (H_a–H_e) at low LSR concentrations (i.e., $\rho \leq 0.6$; see Figure 4).

* (fod) = 1,1,1,2,2,3,3-heptafluoro-7,7-dimethyl-4,6-octanedionato chelate.

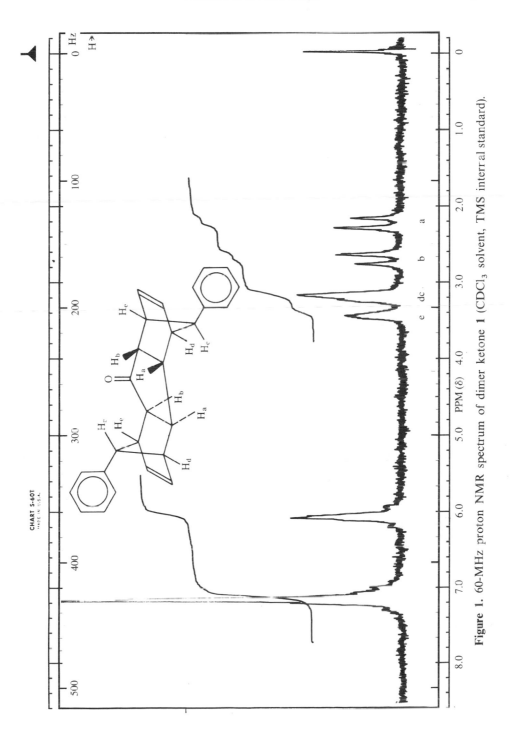

Figure 1. 60-MHz proton NMR spectrum of dimer ketone **1** (CDCl$_3$ solvent, TMS interral standard).

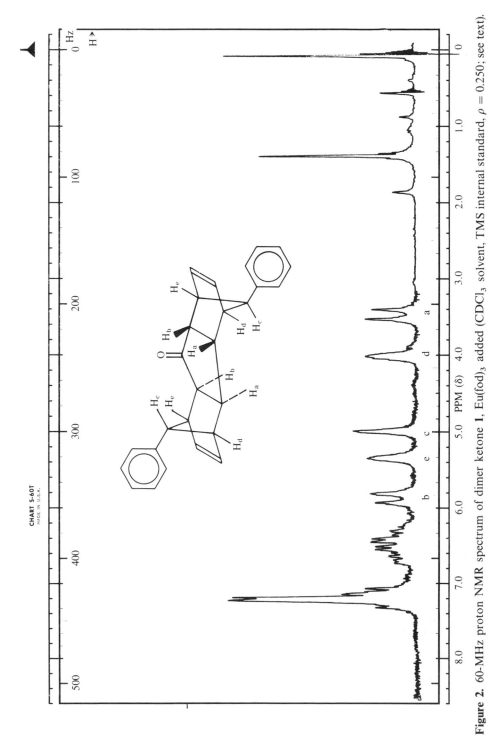

Figure 2. 60-MHz proton NMR spectrum of dimer ketone **1**, Eu(fod)$_3$ added (CDCl$_3$ solvent, TMS internal standard, $\rho = 0.250$; see text).

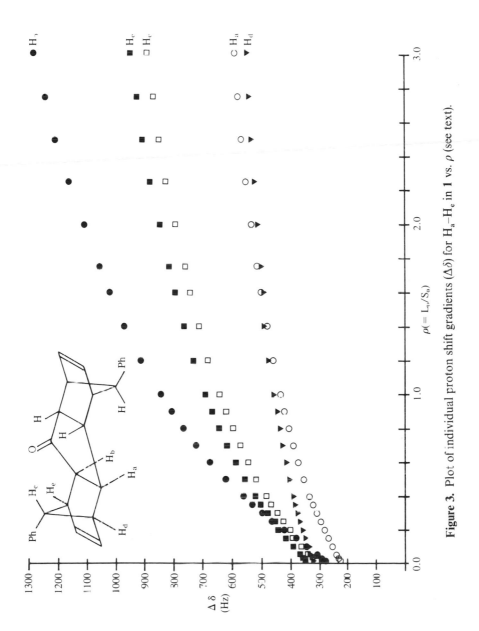

Figure 3. Plot of individual proton shift gradients ($\Delta\delta$) for H_a–H_e in **1** vs. ρ (see text).

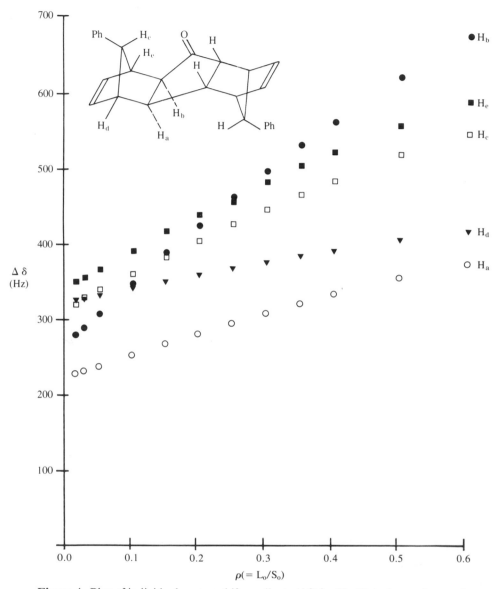

Figure 4. Plot of individual proton shift gradients ($\Delta\delta$) for H_a–H_e in **1** vs. ρ (see text).

A consequence of the rigidity of norbornyl and bicyclo[2.2.2]octyl systems is that their molecular geometries can be defined precisely. Accordingly, LIS data can be used to construct an accurate model for the LSR–solute collision complexes involving these systems.[21–24] From the data shown in Figures 3 and 4, it is possible to construct such a model using bound shift (Δ_i) values for protons H_a–H_e in **1**; the Δ_i values can be calculated using Shapiro and Johnston's LISA2 computer program.[21] In this approach, the position of the europium atom in the LSR is optimized with respect to the substrate using an iterative computer program.[19] In this manner, Δ_i values for protons H_a–H_e can be calculated for each trial position of the europium atom. The trial position is then gradually refined until a best fit is obtained between calculated and experimental Δ_i values for each proton (H_a–H_e) in the substrate. For system **1**, preliminary analysis suggests that the europium atom complexes with the oxygen end of the C=O double bond, colinear with the long axis.[25] Optimal results are obtained for **1** in which the Eu———O=CR$_2$ europium–oxygen bond distance is on the order of 2.0 Å. A similar approach has led to the conclusion that the Pr———O distance is 3.0 Å in the collision complex formed between borneol and *tris*-(2,2,6,6-tetramethylheptane-3,5-dionato) praseodymium[Pr(tmhd)$_3$].[26] In systems that contain one or more symmetry elements, such computer optimization approaches generally have afforded calculated lanthanide–heteroatom distances that are in satisfactory agreement with results of X-ray crystallographic studies.[11,27] This analysis of LSR–substrate complexation is based upon a "one-site model" which assumes that the LSR is coordinated to a single site on the carbonyl oxygen atom, resulting in the formation of a discrete 1 : 1 LSR–substrate complex. Recently, Raber and co-workers[27] have achieved excellent agreement between experimental and calculated LIS values for rigid bicyclic ketones on the basis of a "two-site model" wherein the lanthanide ion coordinates with the carbonyl oxygen at each of the oxygen lone electron pair sites. The observed LIS values are considered to reflect "a weighted time-average of those for each of the two complexes."[27]

Optically active shift reagents, first developed by Whitesides and Lewis,[28,29] have been utilized for direct determination of enantiomeric purity in optically active alcohols and amines.[30–32] Typical of these is *tris*[3-(*tert*-butylhydroxy-methylene)-*d*-camphorato]europium(III) (**2a**), prepared via reaction of 3-*tert*-butylhydroxymethylene-*d*-camphor (**3a**) with europium(III) chloride in the presence of base:[28]

3a: R = C(CH$_3$)$_3$	2a: R = C(CH$_3$)$_3$, M = Eu
3b: R = CF$_3$	2b: R = CF$_3$, M = Eu, Yb
3c: R = C$_3$F$_7$	2c: R = C$_3$F$_7$, M = Eu

Chiral LSRs, such as **3** and other closely related compounds,[30] have been utilized for assigning structure and configuration in a number of rigid bicyclic systems. As an example, consider the effect of added *tris*[(3-trifluoromethyl-hydroxymethylene)-*d*-camphorato]europium(III) (**2b**, M = Eu) on the proton NMR spectrum of the pesticide dieldrin (**4**). Each pair of protons H_2–H_7, H_3–H_6, and H_4–H_5 in (*meso*)-**4** is normally enantiotopic. However, in the presence of chiral LSR **2b**, each pair of protons in **4** becomes diastereotopic and anisochronous; hence, when **2b** is present, coupling between vicinal protons H_4 and H_5 becomes observable (J_{45} = 3.3 Hz).[33]

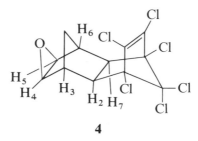

4

[It should be noted that no such coupling would be observed when the NMR spectrum of a racemic (*d*, *l*) mixture is obtained in the presence of added chiral LSR, for corresponding protons in enantiomers would be identical rather than diastereotopic in chiral media.[33]] In similar fashion, the fact that in vivo epimerization of compounds **5** and **6** (both of which are metabolites of dieldrin) proceeded in stereospecific fashion could be proved via analysis of the NMR spectra of these compounds in the presence of chiral LSR **2c**.[34]

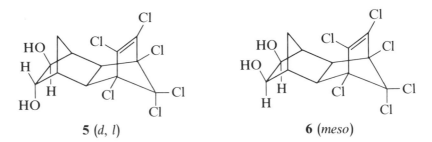

5 (*d*, *l*)   **6** (*meso*)

Intramolecular Nuclear Overhauser Effects

The nuclear Overhauser effect (NOE) phenomenon has been reviewed in a monograph by Noggle and Schirmer,[35] and reviews of chemical applications of NOE have appeared recently.[36–40] Briefly, NOE enhancement of a spin signal is observable when two nuclei are in sufficiently close mutual proximity (3.5 Å or less[36–40]) such that dipole–dipole coupling between them becomes the predominant mechanism for their spin-lattice relaxation (T_1). In such cases, it can be shown[35] that selective saturation of one nucleus (A in a double-irradiation experiment) will result in redistribution of spin populations ("dynamic polarization") in the other (B, singly irradiated) nucleus, with consequent change in the total intensity observed for nucleus B. When the change in total intensity occurs in a nucleus that is in the same molecule as is the spin which is being saturated, this effect is termed an *intramolecular* NOE. The maximum enhancement factor for any intramolecular *homonuclear* NOE experiment is 0.5; the contribution by T_1 relaxation mechanisms other than the dipole–dipole mechanism (T_1^{DD}) will result in an enhancement factor < 0.5. The corresponding maximum factor for a *heteronuclear* NOE experiment is given by $\gamma_A/2\gamma_B$, where γ_A and γ_B are the magnetogyric ratios of the two pertinent nuclei, A and B.[41]

The contribution to T_1 from the intramolecular dipole–dipole interaction between two proximate nuclei varies with the inverse sixth power of the distance r, that separates the two nuclei of interest. A number of investigators took advantage of this fact in assigning stereochemistry and configuration in suitably constructed molecular systems. The first stereochemical application of intramolecular NOE was reported by Anet and Bourn,[42] who studied homonuclear (proton–proton) NOE in system 7. An increase in absorption intensity (ca. 45%, even in the presence of oxygen) of either H_A or H_B in 7 could be observed upon double irradiation of H_B or H_A, respectively, thereby confirming their endo,endo stereochemistry in this half-cage acetate.[42]

exo   endo endo   exo

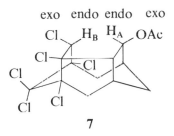

7

In recent years, a relatively small number of papers has appeared in which homonuclear NOE was utilized for the purpose of routine assignment of configuration in substituted norbornanes, bicyclo[2.2.2]octanes, and related systems. Some selected examples that demonstrate this important application of homonuclear NOE appear in Table 1,[45–46] (structures 8–12). The near-maximum value (0.45–0.50) that has been reported for the geminal (6-position) proton–proton NOE enhancement in 2-aryl-2-norbornyl cations (12, Table 1[46]) is noteworthy; this is the first report of an instance wherein NOE enhancement has been observed for geminal methylene protons.

In addition to the intramolecular NOE discussed above, another manifestation of the NOE phenomenon, termed the *generalized Overhauser effect*, has been described.[47–49] In this instance, saturation of one nucleus, A, results in a redistribution of intensity among the individual components of an absorption multiplet corresponding to the nucleus, B, which is being observed. No change in the total intensity of the signal corresponding to nucleus B accompanies this

TABLE 1. Some Stereochemical Applications of Homonuclear Nuclear Overhauser Effects (NOE) in Rigid Bicyclic Systems

Compound[a]	Magnitude of Observed NOE[b]	Reference
8	+0.31	43
9 (R = CN, R′ = phenyl)	+0.08 to 0.09	44
10 (R = CN, R′ = phenyl)	+0.15	44
11 (R = NH$_2$, R′ = phenyl)	+0.16	45
12 (Ar = aryl)	+0.45 to 0.50[c]	46

[a] Nuclei being observed which display significant NOE are circled. Nuclei being double irradiated are indicated by an arrow.
[b] Positive values indicate NOE enhancement of the absorption signal corresponding to the proton being observed.
[c] Measured at − 10°; chemical exchange is slow at this temperature.

effect. Anet[50] has exploited the generalized NOE as an aid to interpreting the proton NMR spectrum of compound **13**.

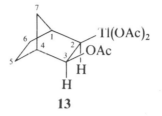

13

In compound **13**, proton coupling to a thallium nucleus ($^{203}Tl_{81}$ and $^{205}Tl_{81}$, spin numbers $I = \frac{1}{2}$, natural abundances 29.50% and 70.50%, respectively) produces a doublet, one line of which corresponds to the $+\frac{1}{2}$ and the other to the $-\frac{1}{2}$ spin states of the thallium nucleus. Because T_1 of thallium is shorter than that of protons, it can be shown that saturation of one of the two absorption signals in this doublet should result in a decrease in the intensity of the other signal.[50] Making use of this phenomenon, Anet[50] was able to (i) assign the various absorption bands in the spectrum of **13** to specific protons in that compounds, and (ii) establish the strong stereochemical dependence of long-range Tl–H couplings in this conformationally rigid molecule.

The usefulness of NOE as an aid to making spectral and configurational assignments in rigid bicyclic systems appears to be well established. In view of the power of the method and the experimental simplicity of its application, it seems somewhat surprising that NOE techniques have not been more extensively exploited for this purpose.

3

STEREOCHEMICAL APPLICATIONS OF NMR CHEMICAL SHIFTS IN RIGID BICYCLIC SYSTEMS

Introduction

The desire to effect the quantitative separation of steric effects from the polar (electronic) effects of substituents has provided a major driving force for the development of structure–reactivity correlations by physical-organic chemists during the preceding four decades[4] (also see refs. 51–54 for background and leading references). Quantitative models for both electronic substituent effects[5,55,56] and steric effects[57] have been developed the applications of which have greatly enhanced our understanding of the energetic factors that govern chemical reaction kinetics and chemical equilibria. The role that nuclear magnetic resonance studies in rigid bicyclic systems related to norbornane and to bicyclo[2.2.2]octane have played in the development of a unified theory of substituent effects[58] is examined in the sections Steric Influences of Substituents on NMR Chemical Shifts and Electronic Influences of Substituents on Chemical Shifts, in this chapter.

It is important to realize at the outset that NMR measurements of relative nuclear shieldings (which relate directly to chemical shifts) are manifestations of magnetic phenomena. These, in turn, relate only indirectly to the many primary factors that contribute to the overall total "screening constant" through which substituents are generally thought to exert their influence upon distant nuclei in the molecule.[51–57] The theory of nuclear shielding originally developed by Saika and Slichter[59] regards the total screening constant, σ, as a summation of individual local screening contributions. It is generally necessary to consider localized and distant paramagnetic and diamagnetic intra- and intermolecular contributions in diagnosing chemical shift differences induced at a given nucleus by substitution for hydrogen elsewhere in the molecule.[60,61]

The foregoing should serve to temper conclusions derived from studies that attempt to interpret observations of substituent-induced differential chemical shifts in terms of a single dominant influence.

Nevertheless, such caveats notwithstanding, numerous attempts have been made to construct suitable molecular systems designed to isolate (or magnify) a clearly dominant contribution to the overall screening constant in such a way that meaningful conclusions regarding the mechanism of transmission of intramolecular substituent effects can be drawn. The choice of nucleus to be observed in the NMR experiment is also critical in this regard. In the following sections the degree to which this approach has met with success in NMR studies in norbornyl, bicyclo[2.2.2]octyl, and related systems is evaluated.

Steric Influences of Substituents on NMR Chemical Shifts

Proton Chemical Shifts

The proton has certainly received more detailed attention than has any other nucleus so far studied by NMR. However, evaluation of the individual contributions to the total screening constant appears to be more complicated in the case of protons than for other nuclei that have been studied. Nevertheless, pronounced proton shifts demonstrably caused by van der Waals (steric) influences have been observed under suitably congested circumstances. An extreme example is provided by the endo,endo-fused alcohols 14 and 15.[62] In systems 14a and 15a (the "O-outside" alcohols), steric compression by the "inside" proton H_B produces a large downfield shift in H_A relative to the chemical shift of the corresponding proton (H_A) in the exo,endo-fused isomer, 16. Similarly, in systems 14b and 15b (the "O-inside" alcohols), the hydroxyl

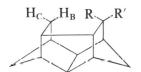

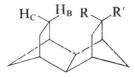

14a: R = H_A, R' = OH
14b: R = OH, R' = H_A

15a: R = H_A, R' = OII
15b: R = OH, R' = H_A

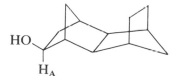

16

oxygen atom deshields H_B by 1.7–2.4 ppm relative to the chemical shift of the normal (uncompressed) endo protons in norbornane.[62] Furthermore, the magnitude of the observed deshielding of H_B was found to vary with the nature of the R group in **14b** and **15b**. The magnitudes of the deshielding effects were observed to decrease in the order $O^-Na^+ > OH > OCH_3 > OBs$; this sequence bears a direct relationship to the order of electron density on the oxygen atom in these substituents.[62] Interestingly, a small shielding effect was noted on H_C in the system **14b** (relative to the chemical shift of the exo protons in norbornane).

Since the pioneering work of Anet and co-workers,[62] numerous examples of "steric deshielding" of protons in sterically congested systems have been reported. Several such examples are presented in Table 2. It seems clear that the data in Table 2 support the qualitative notion that a substantial paramagnetic contribution to the total shielding of a proton in a sterically crowded environment results from steric compression. However, quantitative (or even semi-quantitative) estimation of this contribution is obfuscated by the fact that the magnitudes of proton chemical shifts are the smallest known among nuclei studied actively by NMR. Accordingly, proton chemical shifts are relatively insensitive to environmental factors, and those individual factors which contribute to total proton shielding are often separated and evaluated only with some difficulty[58] (viz. the foregoing caveat).

TABLE 2. CHEMICAL SHIFTS OF STERICALLY COMPRESSED BRIDGE PROTONS IN SYSTEMS RELATED
TO BICYCLO[2.2.1]HEPTANE

Compound	δ_{H_x} (ppm)	δ_{H_y} (ppm)	$\Delta\delta$ (ppm)	Solvent	References
	1.97[a]	0.48	1.49	CCl_4	63, 64
	2.55[a] 2.55	0.97[a] 1.00	1.58 1.55	CCl_4 CCl_4	63–65 63–65
	1.67 1.58	1.35 1.32	0.32 0.26	CCl_4 $CDCl_3$	63 66
	1.06	1.28	0.22	$CDCl_3$	67

TABLE 2 (*continued*)

TABLE 2 (*continued*)

Compound	δ_{H_x} (ppm)	δ_{H_y} (ppm)	$\Delta\delta$ (ppm)	Solvent	References
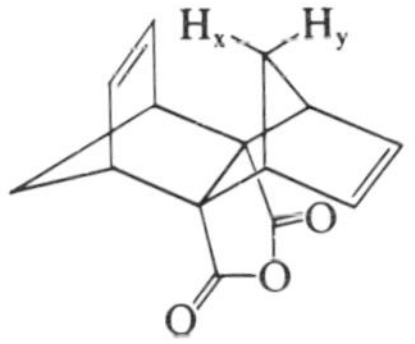 (structure with H_x, H_y; anhydride)	3.00	2.54	1.46	$CDCl_3$	68
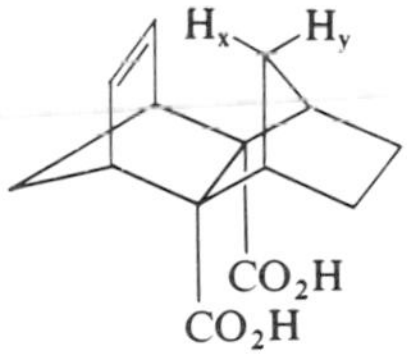 (structure with H_x, H_y; CO_2H, CO_2H)	3.05	0.96	2.09	$(CD_3)_2CO$	68
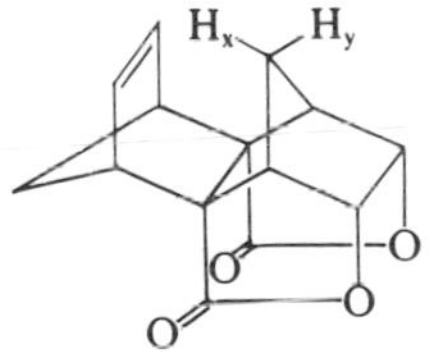 (structure with H_x, H_y; lactone)	2.24	1.58	0.66	$(CD_3)_2CO$	68
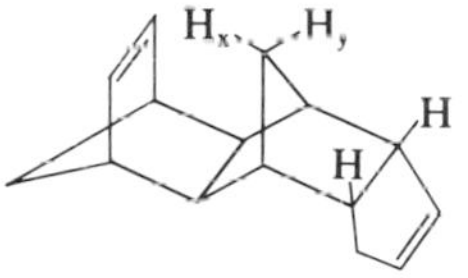 (structure with H_x, H_y; H, H)	2.24	0.75	1.49	$CDCl_3$	69
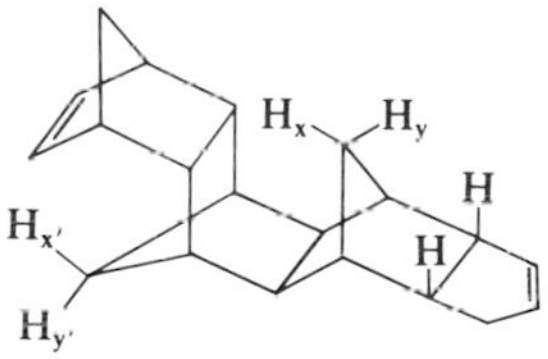 (structure with H_x, H_y, $H_{x'}$, $H_{y'}$; H, H)	2.15 (H_x) 2.28 ($H_{y'}$)	1.16 (H_y) 0.46 ($H_{y'}$)	0.99 1.82	$CDCl_3$ $CDCl_3$	69 69
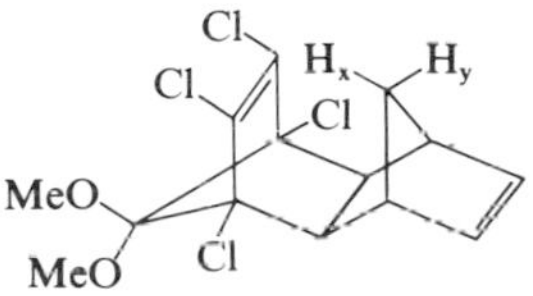 (structure with Cl, Cl, Cl, H_x, H_y, Cl; MeO, MeO)	1.56	1.23	0.33	CCl_4	70
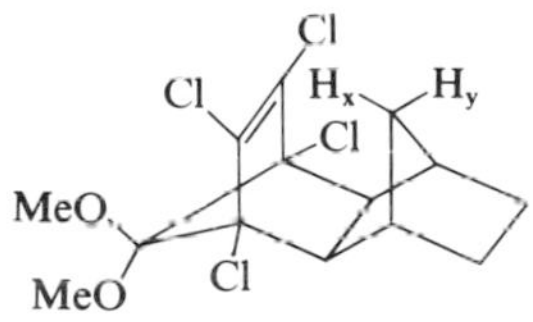 (structure with Cl, Cl, Cl, H_x, H_y, Cl; MeO, MeO)	1.55	0.82	0.67	CCl_4	70

TABLE 2 (*continued*)

TABLE 2 (*continued*)

Compound	δ_{H_x} (ppm)	δ_{H_y} (ppm)	$\Delta\delta$ (ppm)	Solvent	References
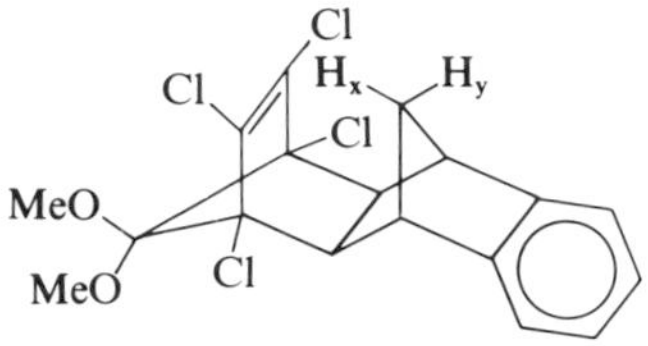	1.85	1.48	0.37	CCl_4	70
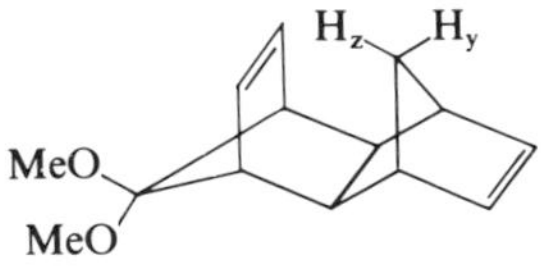	2.55	0.89	1.66	CCl_4	70
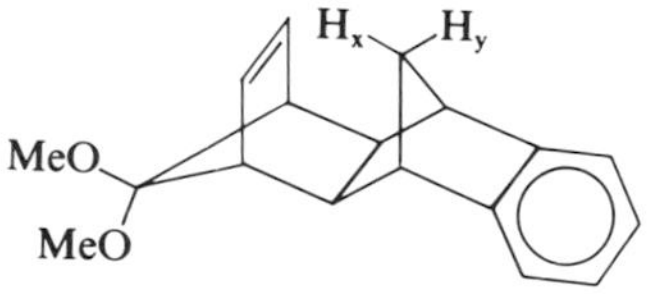	2.85	1.20	1.65	CCl_4	70
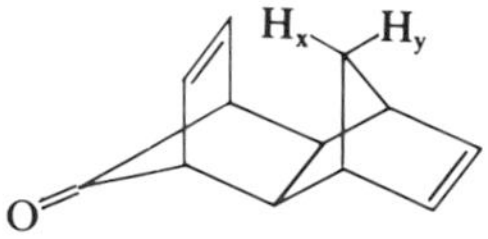	2.46	0.94	1.50	CCl_4	70
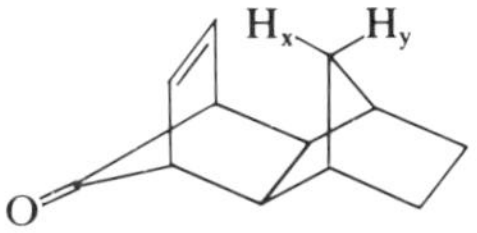	2.08	0.74	1.34	CCl_4	70
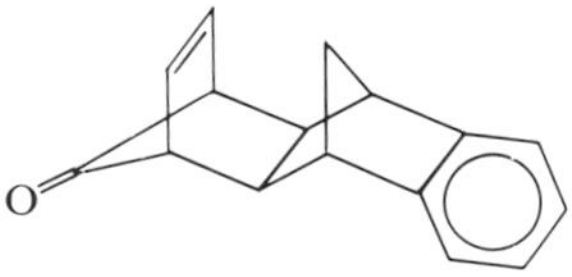	2.66	1.24	1.42	CCl_4	70
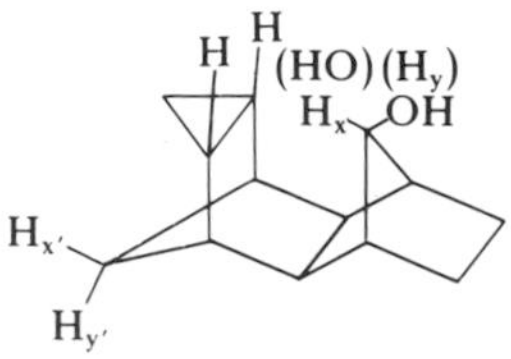	4.98 (H_x)	6.26 (H_y)	-1.28^b	$CDCl_3$	64
	9.05 ($H_{x'}$)	9.51 ($H_{y'}$)	-0.46^b	$CDCl_3$	64

TABLE 2 (*continued*)

TABLE 2 (*continued*)

Compound	δ_{H_x} (ppm)	δ_{H_y} (ppm)	$\Delta\delta$ (ppm)	Solvent	References
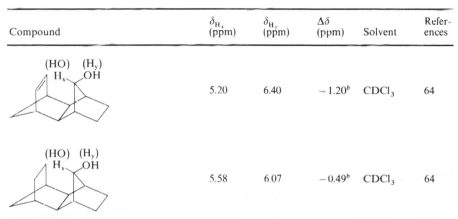	5.20	6.40	-1.20^b	CDCl$_3$	64
	5.58	6.07	-0.49^b	CDCl$_3$	64

[a] These assignments were reversed in the original paper[63]; see ref. 64 for a discussion of the evidence upon which reassignment is based.
[b] $\Delta\delta$ values between bridge protons in these compounds were accounted for in terms of neighboring group anisotropy effects or steric compression effects or a combination of both factors.[64]

Fluorine-19 Chemical Shifts

Intramolecular shielding of the ^{19}F nucleus (like that of ^{13}C and unlike that of ^{1}H) appears to be dominated by the paramagnetic term of the Ramsey equation.[71-74] Because the paramagnetic contribution to overall ^{19}F shielding can be shown to be directly related to the polarity of the C—F bond, considerable effort has been expended to utilize ^{19}F chemical shift data as a probe for elucidating the mechanisms of transmission of electronic substituent effects[4] (see Electronic Influences of Substituents on Chemical Shifts). Fluorine nuclear shieldings are generally much more sensitive to electronic environment than are corresponding ^{1}H shieldings, with the consequence that environmental effects on ^{19}F chemical shifts tend to be greatly magnified relative to the corresponding ^{1}H shift effects.

The availability of exo,endo-fused fluorinated dimethanonaphthalenes (related to the systems shown in Table 2) whose molecular geometries are known accurately would make possible a direct comparison between the effects of steric compression on ^{19}F chemical shifts with the corresponding effects on ^{1}H shifts (discussed in the preceding section). Unfortunately, although the Diels-Alder reaction of hexafluorocyclopentadiene with norbornadiene has been reported[75] to afford two dimethanonapthhalenes (ratio 68 : 32), the configurations of these adducts have not been determined.

Nevertheless, some insight into the effects of steric compression on ^{19}F chemical shifts can be gained through consideration of ^{19}F NMR spectra of other, related alicyclic systems. An example in this regard is provided by *exo*-3,3-difluorotricyclo[3.2.1.0^{2,4}]oct-6-ene (**17**), first reported by Jefford and co-workers.[76] Inspection of molecular models suggest that proton H_{8x} and fluorine F_{3x} nuclei in **17** are in sufficiently close proximity that mutual steric

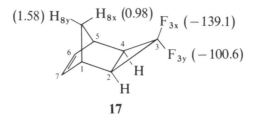

17

compression may conceivably be an important contributing factor in determining their respective chemical shifts (relative to H_{8y} and F_{3y}, respectively). The chemical shift values for H_{8x} and H_{8y} (in parts per million, downfield from internal tetramethylsilane) and F_{3x} and F_{3y} (in parts per million, upfield from internal $CFCl_3$) appear in parentheses beside the corresponding nuclei in structure **17**. Interestingly, the chemical shift of the more sterically crowded proton H_{8x} is seen to be upfield (shielded) relative to H_{8y}; similarly, F_{3x} is shielded relative to F_{3y}.

In addition to these results, Ando and co-workers[77] have reported the ^{19}F chemical shifts (in parts per million upfield from external trifluoroacetic acid reference, CCl_4 solvent) of the bridge fluorine atoms in *syn*- and *anti*-7-fluoro-2-oxanorcarane (**18a** and **18b**, respectively) as being -165.0 and -135.1, respectively. Here, the steric interactions that confront the more hindered *syn*-7-F in **18a** lead to a 30-ppm diamagnetic (upfield, low-frequency) shift for this ^{19}F nucleus relative to that for the corresponding anti isomer (**18b**). Similarly, Dewar and Squires[78] have found that the (sterically compressed) axial fluorine atom in 2,2-difluoro-*trans*-decalin is shielded by 12.5 ppm (acetone solvent) relative to the equatorial fluorine atom in this system.

18a: X = F, Y = H
18b: X = H, Y = F

Other examples wherein steric compression leads to diamagnetic shifts in ^{19}F NMR spectra of fluorinated alicyclic systems have been reported as well.[79] However, the situation is certainly more complicated than this implied generality. Roberts and co-workers[80,81] have reported ^{19}F chemical shifts in a series of monomethyl-2,2-difluoronorbornanes. In structures **20** and **21**, the chemical shifts (in parts per million, upfield from $CFCl_3$) are indicated in parentheses for the *exo*-2 and *endo*-2 fluorine atoms in *syn*-7-methyl-2,2-difluoronorbornane and in *endo*-6-methyl-2,2-difluoronorbornane, respectively. The corresponding

^{19}F chemical shifts for the parent (nonmethylated) 2,2-difluoronorbornane accompany structure **19**. The *syn*-7 methyl group is observed to deshield the *exo*-2 fluorine atom in **20** by 4.8 ppm (relative to the corresponding fluorine atom in **19**), whereas the *endo*-6 methyl deshields the *endo*-2 fluorine atom in **21** by 7.5 ppm (again, relative to **19**). The fact that introduction of a methyl substituent in a 1,3-diaxial relationship to a ^{19}F nucleus in fluorocyclohexane and fluorodecalin systems causes a downfield shift to the ^{19}F nucleus relative to the corresponding system in which this methyl group is absent has been noted.[81-83]

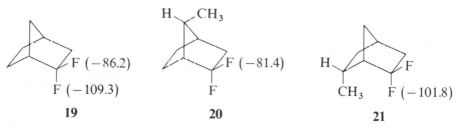

Interestingly, Dence and Roberts [81] have reported their observation of an important vicinal orientation effect upon ^{19}F shielding in methyl-substituted 2,2-difluoronorbornanes. A large ^{19}F diamagnetic shift was found to occur when the dihedral angle between the vicinal carbon–methyl and carbon–fluorine bonds in 0°. The effect is diminished when the dihedral angle is expanded to 60°, and the effect essentially vanishes when the dihedral angle becomes 120°. The observed effects were considered to be significantly larger than could be rationalized on the basis of approximate molecular orbital calculations.[81]

In another study, comparison of ^{19}F chemical shifts along a series of 3-substituted 2,4-dimethylfluorobenzenes with the corresponding shifts in meta-substituted fluorobenzenes was reported.[84] It was found that steric compression of ^{19}F nuclei in aromatic fluorides led to significant downfield ^{19}F chemical shifts (i.e., steric compression results in deshielding of the ^{19}F nuclei in the former system). Representative data are shown in Table 3. Here, the shifts are reported as "substituent chemical shifts" (SCS) [i.e., as differential chemical shifts of the substituted compounds relative to those in the parent (unsubstituted) compounds, fluorobenzene and 2,4-dimethylfluorobenzene, respectively]. In all cases, the SCS value for a given substituent in the more sterically compressed 3-substituted 2,4-dimethylfluorobenzene series is more negative (i.e., results in greater ^{19}F deshielding) than is the corresponding SCS value in the meta-substituted fluorobenzene series.

The contrast observed between results obtained for ^{19}F chemical shifts in the various fluorinated alicyclic and aromatic systems discussed above, each containing a sterically compressed ^{19}F—C bond, points to the general need for further clarification of the effects of steric compression on ^{19}F NMR chemical shifts. However, it should be noted that the examples cited in references 76–78 directly compare the chemical shifts of hindered vs. unhindered C—F bonds.

TABLE 3. 19F Substituent Chemical Shifts (SCS)[a] in m-Substituted Fluorobenzenes and in 3-Substituented-2,4-dimethylfluorobenzenes[b]

Substituent (X)	SCS[c]	SCS[d]
NH$_2$	+0.4	−2.06
N(CH$_3$)$_2$	−0.1	−4.61
OH	−1.3	−3.46
OCH$_3$	−1.1	−4.47
I	−2.4	−9.31
H	0[e]	0[e]
CN	−2.75	−4.27
NO$_2$	−3.45	−5.76

[a] See text for definition of SCS.
[b] Shifts in ppm measured relative to internal 1,1,2,2-tetrachloro-3,3,4,4-tetrafluorocyclobutane (CCl$_4$ solvent). SCS values are positive if the chemical shift of the substituted compound is upfield (more shielded) relative to that of the parent (unsubstituted) compound (in this case, fluorobenzene and 2,4-dimethylfluorobenzene, respectively).
[c] Data from refs. 85 and 86.
[d] Data taken from ref. 84.
[e] By definition.

This direct comparison does not isolate the steric factor from other contributions to the overall ^{19}F nuclear shielding in the instances being compared, as the two C—F bonds in each series of compounds (exo-endo or syn-anti) are in very different electronic and magnetic environments. The comparisons of the type made between **21** and **19** and between **20** and **19** are much better measures of the steric contribution to overall shielding as "inherent" chemical shift differences between exo-2 and endo-2 fluorine atoms (i.e., those differences in the total nuclear shielding, exclusive of an overriding steric contribution) are taken in account when considering the double differences

$$\Delta\delta = (\delta^{20}_{exo\text{-}2\text{-}F} - \delta^{19}_{exo\text{-}2\text{-}F})$$

and

$$\Delta\delta = (\delta^{21}_{endo\text{-}2\text{-}F} - \delta^{19}_{endo\text{-}2\text{-}F})$$

This same type of double difference comparison was made in the work described in references 81–83 but is lacking in the data taken from references 76–78.

Carbon-13 Chemical Shifts

Nuclear shielding in the case of the ^{13}C nucleus is dominated by the paramagnetic contribution, as has been seen to be the case for the ^{19}F nucleus. The routine availability of high-resolution Fourier transform NMR instrumentation in recent years has enabled investigators to obtain natural abundance ^{13}C NMR spectra of superb quality on small quantities of material with relative ease. Like ^{19}F nuclei, ^{13}C nuclei are extremely sensitive to their electronic environment; this sensitivity is reflected by the very broad range of ^{13}C chemical shifts that has been observed, (over 200 ppm[72]; for ^{13}C NMR reviews see refs. 87–91). These factors, coupled with the many desirable structural features displayed by rigid bicyclic systems related to norbornane and bicyclo[2.2.2]octane, have led to extensive, systematic applications of ^{13}C NMR spectroscopy to solving stereochemical problems in these systems.[91]

The effects of steric crowding on ^{13}C chemical shifts in rigid bicyclic systems have been extensively studied. In several instances, ^{13}C chemical shift assignments in these systems have been based upon analysis of the anticipated effects of steric crowding upon individual ^{13}C nuclei.[80,92–100]

For purposes of discussion, it is convenient to analyze substituent steric effects on ^{13}C chemical shifts in terms of α, β, γ, and δ effects (see structures **22** and **23**), where the Greek letter designations refer to the location of the substituent relative to the ^{13}C nucleus that experiences the steric perturbation. In the norbornyl system, steric compression effects of the substituent, X, on ^{13}C chemical shifts can be expected to be most important in the situations depicted in Table 4, (structures **24 29**).

22   **23**

The magnitudes of α- and β-substituent effects on ^{13}C shifts show only a very minor dependence upon substituent orientation, but γ effects display a pronounced sensitivity to the orientation of the substituent.[80] This can be illustrated by considering the ^{13}C chemical shift data obtained for a series of methyl-substituted norbornanes (Scheme 1). In the case of 2-methylnorbornanes (**30** and **31**), the chemical shifts ($\Delta\delta$) at C-3 (β position) are nearly insensitive to the orientation of the 2-methyl group (exo or endo). Even the α carbon (C-2) chemical shifts are very similar for **30** and **31**. However, the γ effects display a marked dependence upon substituent orientation: C-7 is deshielded in **30** (relative to the parent unsubstituted norbornane, presumably because of the operation of a γ-steric effect of the type indicated by structure **26** in Table 4). No such deshielding effect at C-7 is noted in the corresponding endo-2-methyl compound (**31**). Similarly, a strongly deshielding γ-steric effect is evident at C-6 in **31** (see structure **27** in Table 4) but not at C-6 in **30**. Corresponding deshielding γ effects of the type **28** (Table 4) are evident at C-2 and C-3 in **33** (i.e., at the carbon atoms in the ethano bridge that is syn to the 7-methyl group in **33**). Combinations of electronic and steric substituent effects

TABLE 4. Designations of Steric Compression Effects on ^{13}C Chemical Shifts in Substituted Norbornanes

Position of Substituent, X	Position of Reference ^{13}C Nucleus	Type of Effect	Designation[a]
exo-2	C-3	β	24
endo-2	C-3	β	25
exo-2	C-7	γ	26
endo-2	C-6	γ	27
syn-7	C-2, C-3	γ	28
endo-2	C-5	δ	29

[a] The reference ^{13}C nucleus is indicated by a black dot.

Scheme 1. *Substituent Effects on* ^{13}C *Chemical Shifts* $(\Delta\delta_C)^a$ *in Some Methyl-Substituted Norbornanes*[80]

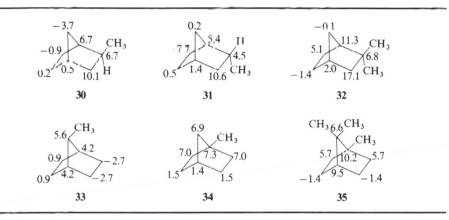

^{a} $\Delta\delta_C = (\delta_C^{RX} - \delta_C^{RH})$.

have been employed to rationalize the corresponding $\Delta\delta$ values in the remaining compounds (**32**, **34**, and **35**)[80].

The effects of steric compression on γ-^{13}C chemical shifts in substituted bicyclo[2.2.2]octanes[101-105] have also been reported. The γ effects in methyl-substituted bicyclo[2.2.2]octanes (Scheme 2[101]) are comparable to the corresponding γ-substituent effects in methyl-substituted norbornanes (Scheme 1[80]), indicating that ring strain has little effect upon the magnitudes of these shieldings. Similar conclusions derive from studies of γ-^{13}C shifts in methyl-substituted norbornenes (Scheme 3[106]) and bicyclo[2.2.2]octenes (Scheme 4[106]).

Scheme 2. *Substituent Effects on* ^{13}C *Chemical Shifts* $(\Delta\delta_C)^a$ *in Some Methyl-Substituted Bicyclo[2.2.2]octanes*

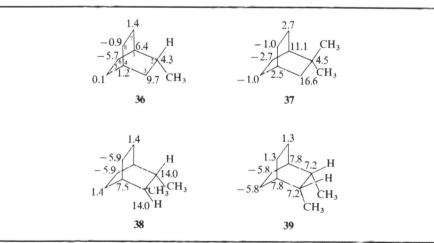

^{a} $\Delta\delta_C = (\delta_C^{RX} - \delta_C^{RH})$.

Scheme 3. *Substituent Effects on* ^{13}C *Chemical Shifts* $(\Delta\delta_C)^a$ *in Some Methyl-Substituted Norbornenes*

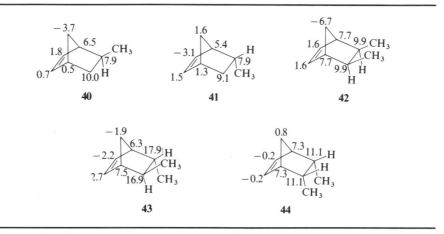

$^a \Delta\delta_C = (\delta_C^{RX} - \delta_C^{RH})$.

Scheme 4. *Substituent Effects on* ^{13}C *Chemical Shifts* $(\Delta\delta_C)^a$ *in Some Methyl-Substituted Bicyclo[2.2.2]octenes*

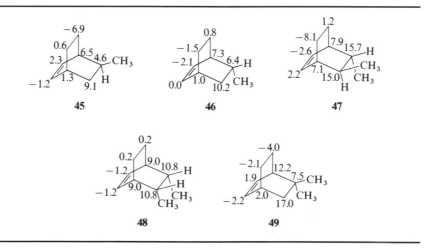

$^a \Delta\delta_C = (\delta_C^{RX} - \delta_C^{RH})$.

From these and many other examples[80] the impression may be gained that steric compression invariably leads to shielding of the affected ^{13}C nucleus. Whereas this appears generally to be the case for steric interactions that take place over three chemical bonds (γ effects), it certainly is not universally true. Indeed, in certain systems, steric compression between substituents and carbon atoms that are mutually separated by four chemical bonds (δ effects) has been shown to result in deshielding of the affected δ-^{13}C nuclei.[107-109]

The δ-^{13}C effect that results from nonbonded steric interactions between *endo*-2-CH$_3$ and *endo*-5-H (as shown in structure **29**, Table 4) is *endo*-2-methylnorbornane does not appear to be significant (cf. $\Delta\delta_C$ for C-5 in structure **31**[101]). However, a dramatic effect is seen for $\Delta\delta_C$ of the 2-methyl group in *endo*-2-hydroxy-*endo*-2-methylnorbornane (**50**), for which

$$\Delta\delta_C^{Me} = [\delta_C^{Me}(ROH) - \delta_C^{Me} \text{ (parent)}] = 2.0$$

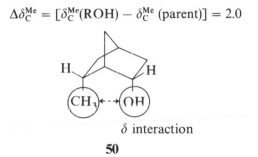

δ interaction

50

(i.e., the δ-steric interaction between the methyl and hydroxyl groups results in a downfield shift of 2.0 ppm for the methyl group in **50** relative to that of the corresponding methyl group in *endo*-2-methylnorbornane).[107] Similar deshielding δ-steric effects have been observed to result from *syn-axial* CH$_3$———OH interactions in some 10-methyl-*trans*-decalols[108] and from *syn*-X—— H$_\delta$ interactions in 2-adamantanol (X = OH)[109] and in 2-adamantanethiol (X = SH).[109]

It is now apparent that the effects of steric crowding over three bonds (γ-steric effects) on ^{13}C chemical shifts are indeed opposites of the corresponding effects over four bonds (δ-steric effects); the former are shielding, whereas the latter are generally deshielding. Accordingly, caution must be exercised when interpreting small shielding differences in the ^{13}C NMR spectra of complex systems.*[107]

Chemical Shifts of Other Nuclei

In addition to the examples discussed in the preceding three sections, the effects of steric compression on ^{31}P chemical shifts have also been studied systematically. Norbornanes and norbornenes bearing phosphorus-containing substituents have been employed in these studies.[111,112] The observation of anticipated γ-steric effects on ^{13}C shifts proved to be valuable in assigning ^{13}C chemical shifts in the NMR spectra of *syn*- and *anti*-7-norbornenylphosphonous dichlorides.[111]

Littlefield and Quin[112] have measured ^{31}P chemical shifts in a number of norbornanes and norbornenes bearing phosphorus-containing substituents in

---

* Results of recent calculations based on the modified INDO finite perturbation theory of ^{13}C chemical shifts suggest that the mechanism of the γ-substituent effect "may be considerably more complex than the popularly accepted C—H bond polarization by nonbonded H———H interactions."[110] Molecular mechanics calculations have been used to analyze sterically induced substituent effects on ^{13}C NMR chemical shifts in substituted norbornanes.[100]

the 7 position (see structures **51–53**, substituents **a–d** in Table 5). These investigators attempted to explain the differences observed for ^{31}P chemical shifts between syn-anti pairs (**52** and **53**, Table 5) in terms of the reduced steric crowding in moving from the anti (**52**) to the syn (**53**) configuration and in terms of the anticipated magnetic anisotropy effect of the 2,3 double bond when the substituent is syn to the double bond (**53**) vis-à-vis the corresponding system in the anti configuration (**52**). The situation under consideration is far from being straightforward. This can be seen by the fact that relief of steric strain in moving from the anti configuration to the less crowded syn configuration produces a net shielding effect on the ^{31}P nucles for X = PMe$_2$, but a net deshielding effect is observed for the ^{31}P nuclei in the other three substituents (X = PCl$_2$, P(S)Me$_2$, and $^+$PMe$_3$ I$^-$, Table 5). These results were also viewed

TABLE 5. ^{31}P NMR CHEMICAL SHIFTS^a IN 7-SUBSTITUTED NORBORNANES AND NORBORNENES[112]^b

	Compound		
Substituent	51	52	53
a: X = PMe$_2$	−62.0	−60.2	−61.3
b: X = PCl$_2$	+199.7	+190.9	+199.7
c: X = P(S)Me$_2$		+30.2	+36.5
d: X = $^+$PMe$_3$ I$^-$		+19.8	+24.2

^{a} CDCl$_3$ solvent; shifts are referred to 85% H$_3$PO$_4$. Positive shifts are upfield of this reference standard.
^{b} Reproduced with permission from *Organic Magnetic Resonance*, Vol. 12, No. 4 (1979), Heyden & Son Ltd, London.

as being inconsistent with Gorenstein's hypothesis,[113] which relates steric compression effects on ^{13}C, ^{19}F, and ^{31}P chemical shifts to anticipated bond angle distortions "which arise from a coupling of bond angles to torsion angles." [113]

The same types of complications are present in the comparison being made between systems **52** and **53** as have been encountered previously when considering ^{19}F chemical shifts in a variety of rigid systems (see discussion in Fluorine-19 Chemical Shifts[76–78]). Once again, it would be desirable to design a system composed of a series of compounds in which the environment of the ^{31}P nucleus was maintained essentially constant, with the exception of the steric compression factor. Perhaps a series of methylated ^{31}P-substituted norbornanes could be employed to advantage much as the methylated systems shown in Schemes 1–4 were used successfully to study the effects of steric compression on ^{13}C chemical shifts.

Electronic Influences of Substituents on NMR Chemical Shifts

In addition to the steric influence that substituents exert on chemical shifts (discussed in the preceding sections), it is generally recognized that substituents also can exert an electronic influence that affects the overall distribution of electron density in a molecule (and at a particular nucleus in that molecule). The electronic effects of substituents manifest themselves through their effect on the chemical shifts of affected, distant nuclei in the molecule. A number of attempts have been made to correlate the simple concept of electronic substituent effects with the physical measurement of NMR chemical shifts.[4,51–56,58,114–117]

Rigid bicyclic systems have been utilized extensively as substrates for these NMR studies. They have served not only to define the geometric (orientation) factors that govern electronic substituent effects, but they have also been utilized to elucidate the detailed mechanisms by which the electronic effects of substituents express themselves. In particular, NMR studies on substituted rigid bicyclic systems have been utilized extensively to further investigate the controversial "through-space" vs. "through-bond" mechanisms of transmission of polar substituent effects (see Fluorine-19 Chemical Shifts, in this chapter).

In recent years, a vast literature has accumulated in which chemical shifts measurements in a great number of molecular systems have been correlated with a variety of "substituent parameters." The basis of such "linear free energy relationships" (LFER) and discussions of their applications to NMR spectroscopy (and vice versa) have been the subjects of a number of recent reviews.[4,51–56,58,114,115] Given its stated focus, the present review accordingly is concerned primarily with detailing those NMR-based studies of electronic substituent effects which have extensively utilized substituted bicyclic systems related to norbornane and bicyclo[2.2.2]octane. In doing so, we seek to emphasize the structural features of these rigid bicyclic systems which offer particular advantage to investigators who seek to elucidate the mechanisms of transmission of electronic substituent effects. Additionally, the usefulness of the resulting substituent effect correlations in assigning stereochemistry in rigid bicyclic systems bearing a variety of substituents will, it is hoped, become evident.

Proton Chemical Shifts

The difficulties attendant with evaluation of individual contributions to overall proton screening have been noted previously (see Steric Influences of Substituents on NMR Chemical Shifts, Proton Chemical Shifts, in this chapter). Nevertheless, some useful correlations of proton chemical shifts with electronic substituent effects have been developed. In the early 1960s, the first observations that permitted the correlation of proton chemical shifts with molecular structure parameters were reported. Fraser[118] utilized the fact that the 5,6-*exo* protons in norbornene derivatives are deshielded relative to the corresponding isomeric 5,6-*endo* ring protons to establish the configuration of these protons (and, hence, of the substituents) in norbornenes bearing substituents in the 5 and/or 6 ring positions. The difference between exo and endo proton chemical shifts in norbornenes was ascribed to the magnetic anisotropy of the 2,3 double bond in this system. Similarly, exo ring protons have been observed to be deshielded relative to the corresponding endo protons in 2- and 2,3-substituted norbornanes as well.*[120] Extension of this argument led Tori and co-workers[122] to suggest that this same factor would be expected to shield the bridge proton that was syn to the double bond (H_{7s} in **54**) relative to the corresponding anti bridge proton (H_{7a} in **54**). However, results obtained from

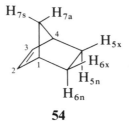

54

subsequent studies involving both double-resonance experiments[123,124] and analysis of the proton NMR spectra of specifically deuterated norbornenes[63,125] required that Tori's original bridge proton assignments in norbornene[122] be reversed. Tori's group, in their reinvestigation of the question of norbornene[123] and bicyclo[2.2.2]octene[124] relative bridge proton chemical shifts, cited the possible significance of unusual magnetic shielding effects and/or novel electronic effects in accounting for the relative chemical shifts of these bridge protons. Further refinement of this explanation was suggested by Franzus and co-workers,[125] who forwarded a carefully reasoned geometric argument to account for the observed H_{7s}–H_{7a} chemical shift difference in norbornene. Additional results reported by Tori's group[126] which suggest that the chemical shift difference $\Delta\delta = (\delta_{H_{7s}} - \delta_{H_{7a}})$ in a series of ring-substituted benzonorbornenes (**55**) remains invariant over a wide range of substituents, X,

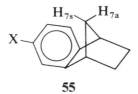

55

---

* Exo ring protons in unsubstituted norbornane are deshielded by ca. 0.3 ppm relative to the endo ring protons in this system.[121]

lend support to the geometric interpretation[125] over an alternative electronic interpretation in this regard. Taken as a group, these studies emphasize the hazards concomitant with attempts to account for small observed differences in proton chemical shifts on the basis of anticipated magnetic anisotropy effects of distant substituents!

Important progress in successfully correlating substituent effects on proton chemical shifts was garnered by Williamson and co-workers in 1963–1964.[127-129] Prior work by Dailey et al.[130,131] had established that internal chemical shifts in substituted ethanes, CH_3CH_2X, $\Delta\delta = (\delta_{CH_3} - \delta_{CH_2})$, could be correlated linearly with the electronegativities of the substituents, X, in this system. Subsequently, Williamson[127] found that the chemical shifts of the exo-6 and endo-6 ring protons in a series of endo-5-substituted 1,2,3,4,7,7-hexa-chloronorbornenes (H_{6x} and H_{6n}, respectively, in 56) could be correlated linearly with substituent electronegativity. Extension of this work to include a study of the proton NMR spectra of a series of para-substituted endo-5-aryl-1,2,3,4,7,7-hexachloronorbornenes (57) revealed that proton chemical shift data

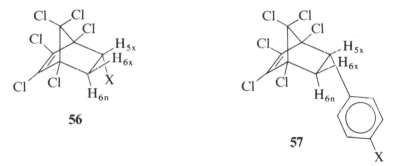

could be correlated linearly with Hammett σ constants[51,52,132] for the various substituents, X.[128] This important observation confirmed the applicability of LFERs to correlating proton chemical shift data in alicyclic systems. Hitherto, such correlations had been restricted to studies of substituent effects on chemical shifts of protons directly bonded to an aromatic (benzene) ring.[4,133,134] Similar linear correlations of H_{7s} and H_{7a} chemical shifts in substituted benzo-norbornadienes (55) and benzonorbornenes with Hammett σ were later reported by Tori's group.[126]

The important observations discussed above[126-129] suggest that the electronic effects of substituents (rather than their magnetic anisotropic effects) are the major factors responsible for determining the overall substituent effect on proton chemical shifts in the systems studied.* However, these early observations are purely empirical: They do not serve to define either the exact nature of these "electronic effects"; neither do they define the mechanisms of trans-

---

* However, the effects of carbonyl-containing substituents on proton chemical shifts in substituted bicyclo[2.2.2]octanes have been rationalized in terms of the magnetic anisotropy of the carbonyl group.[135]

mission of these effects from the substituent to distant proton nuclei in the molecule. The original Hammett σ values are known to be composed to polar and resonance (also steric!) contributions to the overall "substituent effect."[4,5,54,55] Because the approach utilized by Williamson[127,128] and by Tori and co-workers[126] relies on correlations of proton chemical shifts with Hammett σ, it seems clear that, in the absence of steric contributions, no separation of polar contributions from resonance contributions to the overall electronic substituent effect can result from these studies. However, considerable progress in this direction has been garnered through consideration of electronic substituent effects on ^{19}F and ^{13}C chemical shifts in rigid bicyclic systems. A brief account of these investigations appears in the following sections.

Fluorine-19 Chemical Shifts

Pioneering work by Taft and co-workers,[85,86,114,136–142] later elaborated upon by other investigators,[4,143–152] established the utility of ^{19}F NMR chemical shifts as a probe for elucidating the nature of electronic substituent effects and their modes of transmission in aromatic molecules. The difficulties encountered when attempting to separate polar and resonance contributions to overall electronic substituent effects in aromatic molecules led investigators to seek other molecular systems with which to accomplish this purpose. Rigid bicyclic systems seemed ideal for use in this regard. Their relatively fixed geometries obviate the uncertainties attendant upon estimating internuclear distances in nonrigid systems. Equally important, saturated systems such as norbornane and bicyclo[2.2.2]octane lack the conjugated pi-electron systems that are necessary to the transmission of resonance effects and that, accordingly, generally obfuscate interpretations of electronic substituent effects in aromatic systems.

Initial investigations by Dewar and co-workers[143–145] involved correlation of electronic substituent effects on ^{19}F chemical shifts in aromatic systems utilizing a two-parameter "F-M"[153] approach. In developing this method, it was shown that substituents exert two different effects in aromatic systems: a pi-electron polarization effect and a direct, through-space electrostatic field effect on the distant ^{19}F nucleus. The classical (Branch-Calvin) through-bond "inductive effect" of the substituent was neglected in Dewar's approach. Subsequent ^{19}F NMR studies by Dewar and Squires[78] on fully saturated fluorinated systems (e.g., geminally difluorinated *trans*-decalins and steroids) revealed that a polar substituent exerted its electrostatic (field) effect at a distant fluorine atom directly rather than via σ-electron polarization of the carbon–fluorine bond.

An opposing view, developed by Taft and co-workers,[136–142] utilizes a dual substituent parameter (DSP) equation to correlate substituent effects on ^{19}F chemical shifts in aromatic molecules. Taft's approach takes into consideration both resonance and σ-inductive effects of substituents, and it attempts to separate Hammett substituent constants (σ) into inductive (σ_I) and resonance (σ_R) contributions.*

Despite lingering differences, many fundamental aspects of the two opposing views discussed above have been reconciled.[56,149,155] Nevertheless, it is generally recognized that additional clarification of the relative contributions of direct field effects and pi-electron related effects of substituents on ^{19}F chemical shifts in aromatic systems is much needed.[56,135] Substituted, fluorine-containing rigid bicyclic systems have been extensively utilized for this purpose (e.g., systems 58–62, Table 6). The observations reported in these studies and the conclusions derived therefrom are summarized in Table 6. Together, they provide insight into the overriding complexities associated with attempts to clarify the nature of electronic substituent effects and the mechanisms by which they are transmitted in molecular systems. Additional clarification of these important questions has been gleaned through investigation of substituent effects on ^{13}C chemical shifts in substituted bicyclic systems. A brief account of these studies appears in the following section.

Carbon-13 Chemical Shifts

As pointed out earlier, both ^{13}C and ^{19}F nuclear shielding are dominated by the paramagnetic term of the Ramsey equation. Accordingly, it is not surprising that ^{13}C NMR shift studies have relied upon the same kinds of molecular systems and theoretical approaches that have been utilized to correlate substituent effects on ^{19}F chemical shifts. The use of ^{13}C shifts for this purpose offers certain advantages vis-à-vis ^{19}F shift studies. For example, modern NMR instrumentation has enabled investigators to obtain natural abundance ^{13}C NMR spectra routinely. Therefore, no extraordinary synthetic strategies are required to produce the systems needed in connection with studies of substituent effects on ^{13}C NMR spectral shifts.

Some representative examples that are concerned with analyzing electronic substituent effects on ^{13}C chemical shifts in rigid bicyclic systems are presented in Table 7. In general, conclusions drawn from these investigations complement and extend those which have been derived from corresponding ^{19}F chemical shifts studies on substituted, fluorine-containing rigid bicyclic systems.

Several conclusions that emerge from these ^{13}C NMR shift studies merit consideration. The pioneering work of Lippmaa and co-workers[93] established the approximate additivity of substituent effects on ^{13}C chemical shifts in rigid bicyclic systems, an observation that has been confirmed by a number of investigators.[87–91] This observation has led to the development of "substituent parameters" that can be used to calculate approximate [13]C

---

* Recent evidence suggests that a dual substituent parameter treatment for analysis of substituent effects on ^{13}C NMR chemical shifts may not be "significantly superior" for a simpler single substituent parameter approach.[154]

TABLE 6. Electronic Substituent Effects on ^{19}F NMR Chemical Shifts in Rigid Bicyclic Systems

System Studied	Observations/Conclusions	References
	Anomalous shielding of a distant ^{19}F nucleus was observed to be induced by X = F and CO_2Et in these systems. The anomaly was explained in terms of substituent-induced structural variations that produced concomitant changes in hybridization and in electron distribution in the molecule. More recent results suggest that ^{19}F substituent chemical shifts in **58** are "essentially a manifestation of electric field and electronegativity effects."[164]	156, 157, 164
	Anomalous shielding of a distant ^{19}F nucleus was observed to be induced by X = F and CO_2Et in these systems. The anomaly was explained in terms of substituent-induced structural variations that produced concomitant changes in hybridization and in electron distribution in the molecule. More recent results suggest that ^{19}F substituent chemical shifts in **58** are "essentially a manifestation of electric field and electronegativity effects."[164]	156, 157, 164
	Anomalous shielding of a distant ^{19}F nucleus was observed to be induced by X = F and CO_2Et in these systems. The anomaly was explained in terms of substituent-induced structural variations that produced concomitant changes in hybridization and in electron distribution in the molecule. More recent results suggest that ^{19}F substituent chemical shifts in **58** are "essentially a manifestation of electric field and electronegativity effects."[164]	156, 157, 164
	Anomalous shielding of a distant ^{19}F nucleus was observed to be induced by X = F and CO_2Et in these systems. The anomaly was explained in terms of substituent-induced structural variations that produced concomitant changes in hybridization and in electron distribution in the molecule. More recent results suggest that ^{19}F substituent chemical shifts in **58** are "essentially a manifestation of electric field and electronegativity effects."[164]	

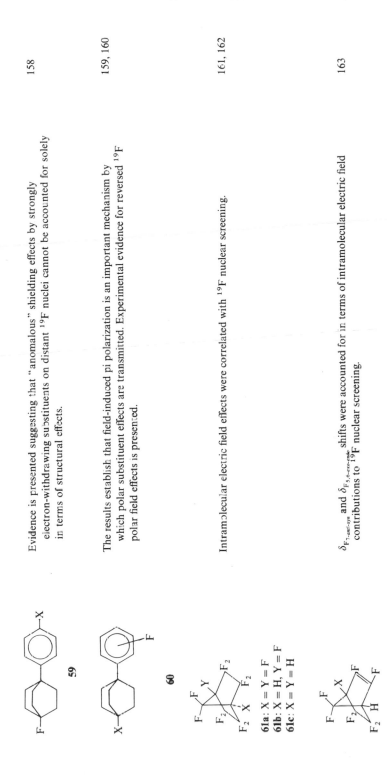

158

Evidence is presented suggesting that "anomalous" shielding effects by strongly electron-withdrawing substituents on distant ^{19}F nuclei cannot be accounted for solely in terms of structural effects.

159, 160

The results establish that field-induced pi polarization is an important mechanism by which polar substituent effects are transmitted. Experimental evidence for reversed ^{19}F polar field effects is presented.

161, 162

Intramolecular electric field effects were correlated with ^{19}F nuclear screening.

163

$\delta_{F7\text{-}anti\text{-}syn}$ and $\delta_{F5,6\text{-}exo\text{-}endo}$ shifts were accounted for in terms of intramolecular electric field contributions to ^{19}F nuclear screening.

59

60

61a: X = Y = F
61b: X = H, Y = F
61c: X = Y = H

62

TABLE 7. ELECTRONIC SUBSTITUENT EFFECTS ON ^{13}C CHEMICAL SHIFTS IN RIGID BICYCLIC SYSTEMS

Systems Studied	Observations/Conclusions	References
(structures)	(i) α- and β-substituent effects consist of a characteristic inductive contribution plus a "variable term which is governed by the state of the C_α—C_β bond.",[80]; (ii) β-substituent effects level off to a maximum (limiting) value with increasing substitution, but α-substituent effects show no sign of leveling off with increasing substitution; (iii) sterically-induced γ-substituent effects are observed in geometrically appropriate instances; (iv) γ-substituent effects resulting from substitution at the *exo*-2-position of norbornane result in ^{13}C shift changes at C-6 that correlate with the electronegativity of the substituent. No such correlation was observed for ^{13}C shift changes at the other two γ carbons (C-4 and C-7).	80
(structures)	(i) ^{13}C shifts are very sensitive to changes in substituent geometry; (ii) substituent influences on ^{13}C chemical shifts are approximately additive. Substituent parameters can be derived empirically that are useful for calculating approximate ^{13}C shifts values in bicyclic systems; (iii) ^{13}C shift differences are caused by 1,4-nonbonded interactions between the ^{13}C nucleus and substituents heavier than hydrogen. Magnetic anisotropy effects of substituents are relatively unimportant in this regard.	93

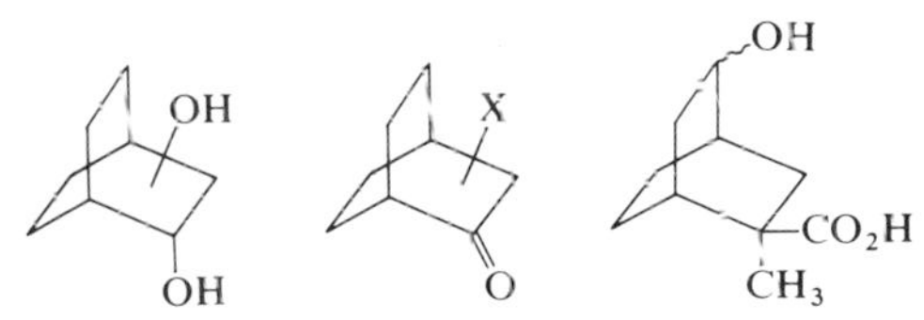

102

Substituent parameters were derived from data obtained for 2-substituted bicyclo[2.2.2]octanes; these parameters were found to be useful for determining the relative stereochemistry and C-2 and C-5.

165

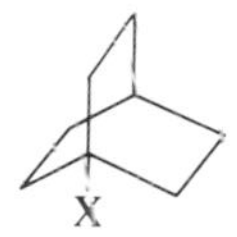

Substituent effects on ^{13}C shifts at C-2 and at C-3 (β and γ positions) were found to decrease with increasing number of intervening carbon–carbon bonds between the ^{13}C nucleus and the substituent. Shieldings at C-4 (δ position) do not follow this relationship.

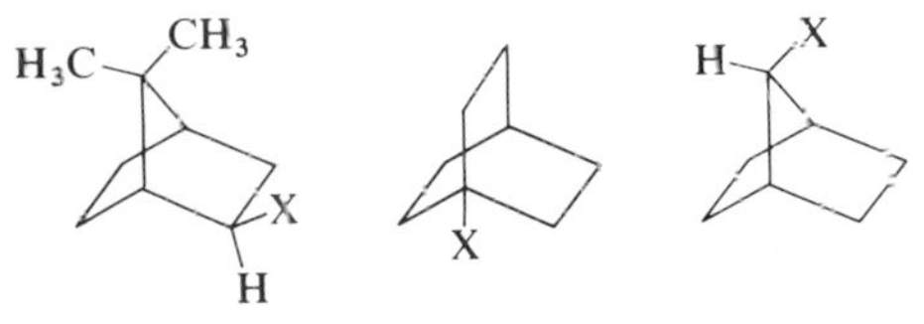

166

Electrostatic field effects of the substituents X, cannot account for the observed substituent effects on the C-4 (δ position) chemical shift.

167

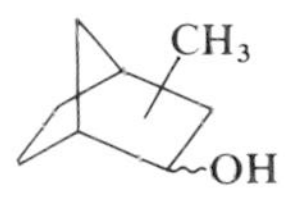

Additivity of electronic substituent effects on ^{13}C chemical shifts was observed.

168, 169

Correlation of ^{13}C shielding data with estimated dihedral angles between vicinal methyl groups was observed.

(X = F, Cl, Br, I)

TABLE 7 (*continued*)

TABLE 7 (*continued*)

Systems Studied	Observations/Conclusions	References
60	System **60** was considered to be capable of isolating polar field effect phenomena (σ_I effect) from other mechanisms by which electronic substituent effects can be propagated. The results establish the importance of field-induced pi polarization as a mechanism of propagation of polar substituent effects.	159, 160
	Substituent effects on ^{13}C shifts could be interpreted solely in terms of the field-induced pi-polarization mechanism; mesomeric effects were considered to be negligible (or effectively constant along the series).	170, 171
(X = H, CH$_3$, CO$_2$H, Br, Cl, NO$_2$)	^{13}C chemical shifts of the low-field carbonyl carbon atoms correlate linearly with σ_I parameters of the substituents X. "The shielding of the carbonyl carbons brought about by electron withdrawing substituents is attributed to a field effect of the substituent which serves to increase the C=O bond order."[172]	172
	^{13}C shifts can be accounted for using Lippmaa and co-workers'[97] substituent shift parameters.	173

174

The additivity of substituent effects on ^{13}C chemical shifts in substituted norbornanes which had been observed earlier,[93] was confirmed in this study.

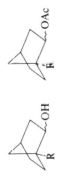

(R = H, CH$_3$, CH$_2$CH$_3$)

175

"α-, β-, and γ-substituent effects in the 1-norbornyl system appear to arise principally from through-bond effects. α-Shieldings are sensitive to strain at C-α but β-shieldings are not sensitive to strain at C-β. γ-Substituent effects at C-3,5 are somewhat influenced by the fact that they do not bear a true antiperiplanar orientation with the substituent at C-1. δ-Substituent effects at C-4 are not purely inductive; a 1,4-field effect or α,γ-hyperconjugative effect is also involved."[175]

(X = exo- and endo-Cl, Br, OH)

176

^{13}C chemical shift assignments for exo- and endo-2-norbornanol[131] were found useful in making chemical shift assignments in exo- and endo-2-halonorbornanes.

TABLE 7 (continued)

TABLE 7 *(continued)*

Systems Studied	Observations/Conclusions	References
(X = CH₃, CH₂OH)	^{13}C shieldings observed for the *trans*-2,3-disubstituted systems are in agreement with the values predicted by simple additivity. This result indicates that significant steric interactions between trans-vicinal substituents are essentially absent (dihedral angle = ca. 120°). *Cis*-vicinally substituted systems exhibit ^{13}C shieldings that deviate from those expected on the basis of simple substituent effect additivity. These deviations can be interpreted "directly in terms of the known geometric dependence of substituent effects."[178]	177
	^{13}C shifts at C-2, C-4, and C-9 in **64** are more shielded than expected on the basis of simple additivity. These extraordinarily high shieldings at C-2 in **64** are rationalized by a through-bond "alternating interaction mechanism."[133] By way of contrast, ^{13}C shifts in **63** are in accord with expectations based upon simple additivity, and they can be rationalized in terms of an electronic through-space (field interaction) mechanism of transmission of substituent effects.	179
(X = H, OH, OCH₃, OAc, Cl, Br, I) and derivatives of these two systems	Linear correlation of ^{13}C chemical shifts with λ_{max} in the ultraviolet absorption spectra of these systems was observed. It was concluded that ^{13}C shifts "are associated with changes in the pi-bond properties of the *s-cis*-butadiene chromaphore."[180]	180

chemical shifts in rigid bicyclic systems.[93] In addition, the importance of substituent orientation on the magnitude of γ-substituent effects has been noted. This factor appears to be important for γ effects, but it appears to be of minor significance in the case of α- and β-substituent effects.[80] However, the origin of these substituent orientation effects is not completely understood; one cannot be certain whether the presence (or, indeed, the absence) of substituent orientation effects is a purely electronic phenomenon or whether instead it mirrors subtle differences in molecular geometry that necessarily accompany changes in substituent orientation.

A significant portion of the ^{13}C NMR work included in Table 7 was originally undertaken in an effort to clarify the nature of electronic substituent effects and, in particular, to elucidate the mechanism by which electronic substituent effects were transmitted to distant sites in the molecule. The approaches that have been used in ^{13}C NMR studies on rigid bicyclic systems parallel those which have been employed in the corresponding ^{19}F NMR studies (discussed in the previous section). It is therefore not surprising to find that conclusions drawn from the ^{13}C NMR studies have aroused controversy similar to that which arises from the corresponding ^{19}F NMR studies. Nevertheless, current understanding of the nature of electronic substituent effects and insight into the mechanisms of their transmission have been advanced materially through pursuit of the kinds of ^{13}C NMR shift studies that appear in Table 7.

It seems to be generally accepted that the use of rigid bicyclic systems permits isolation of polar from resonance contributions to the overall electronic effect of a given substituent. What is not clear is whether this polar substituent effect is transmitted entirely through space (field effect), entirely through the σ-bonding framework (σ-inductive effect), or via some combination of these modes of transmission. Recent calculations lend support to the electronic field effect mechanism; linear correlations of substituent-induced ^{13}C NMR shifts in substituted cyclohexanes with "square electric field effects" $\langle E^2 \rangle$ and with "linear electric field effects" E_Z have been reported.[181] However, Wiberg and co-workers[166] recently have demonstrated that Schneider and Freitag's approach[181] fails to account for either the sign or the magnitude of ^{13}C differential shifts of C-2 (C-3) and C-5 (C-6) in 7-chloronorbornanes and of C-4 in 4-halobicyclo[2.2.2]octanes. Accordingly, Wiberg and co-workers concluded that the field effect is not a major contributor to γ- and δ-substituent effects on ^{13}C chemical shifts in these systems.[166] Instead, the importance of substituent-induced magnetic fields, even at locations as remote as the δ position (relative to the substituent), was stressed; this conclusion was reinforced by the results of a factor analysis study of the relevant ^{13}C chemical shifts.[166]

Despite the extensive efforts that have been expended to clarify the nature and mechanism of transmission of electronic substituent effects via ^{13}C chemical shift studies, these phenomena remain shrouded in mystery. The results cited in Table 7 suggest that no single mechanism suffices to account for overall electronic substituent effects on ^{13}C chemical shifts in rigid bicyclic systems, even in the absence of complicating steric and strain effects. Indeed, a given substituent in a given molecule appears to exert its electronic effects via different mechanisms at different sites in the molecule![80,166,175,179] Nevertheless, consistent patterns of substituent effect behavior have emerged that have permitted investigators to construct linear correlations of substituent effects with ^{13}C chemical shifts. These correlations, in turn, have enabled investigators to make stereochemical assignments in complex, highly substituted bicyclic systems. The successes that have been enjoyed in this regard have demonstrated the viability of the NMR approach. In turn, these should serve to stimulate additional NMR investigations of substituent effects on ^{13}C chemical shifts in rigid bicyclic systems.[182] *

Correlation of NMR Chemical Shifts with Charge Densities in Bicyclic Carbonium Ions

Proton and ^{13}C NMR have been extensively utilized as tools for the elucidation of the detailed structure of carbocations. Inherent in the NMR approach is the assumption that ^1H and ^{13}C chemical shifts in these cationic systems reflect the distribution of charge density in the molecule. It is further assumed that comparison of observed chemical shifts in these systems with the corresponding shifts in model carbocationic systems should be capable of delineating the extent to which delocalization of positive charge from the ionic center occurs in the former systems. In 1973, Olah and Westermann[185] stated that "^{13}C NMR shifts, if used with proper consideration of all factors involved, are a very powerful tool for studying the structure of carbocations, including the trend of charge distribution." This optimistic statement belies the complexity of the problem; a number of the "factors involved" have been assessed recently in critical reviews,[186-189] and the appropriate cautionary note has been sounded. In this section, we consider charge density–NMR chemical shift correlations in rigid bicyclic carbonium ions and the role that these correlations have played in defining the structures of "nonclassical carbonium ions" in these systems. The reader is referred elsewhere[1,190-199] for a complete discussion of the classical vs. nonclassical carbonium ion controversy.

---

* Recently, ^{13}C and ^{19}F NMR studies of a number of bridgehead metalloidal-substituted phenylbicyclo[2.2.2]octyl and (*m*- and *p*-)fluorophenylbicyclo[2.2.2]octyl derivatives have permitted σ_I values to be determined for several metalloidal (e.g., SiX_3, SnX_3, GeX_3, and PbX_3) substituents attached to an sp³-hybridized carbon center.[183] The ^{13}C and ^{19}F NMR spectra of a number of $Cr(CO)_3$ complexes of bridgehead-substituted phenylbicyclo[2.2.2]octyl and (*m*- and *p*-)fluorophenylbicyclo[2.2.2]octyl derivatives have been reported recently. The polar substituent parameter, σ_I, has been found to be more positive for a $Cr(CO)_3$-complexed phenyl substituent ($+0.35$) than for an uncomplexed phenyl substituent ($+0.18$) attached to an sp³-hybridized carbon center. This observation confirms earlier conclusions that $Cr(CO)_3$ coordination enhances the electron-withdrawing capability of a phenyl ring.[184]

The development of low-nucleophilicity "superacid" solvent systems by Olah and his collaborators[198,199] in the early 1960s permitted the direct observation of the norbornyl carbonoium ion by NMR spectroscopy. The low-temperature proton NMR spectrum of this species prepared from exo-2-fluoronorbornane in SbF_5–SO_2 or in SO_2ClF solution was reported by Saunders, Schleyer, and Olah in 1964.[200] At room temperature, the proton NMR spectrum of the norbornyl carbonium ion consists of a single broad absorption signal centered at $\delta 3.75$; this results from complete equilibration of all protons in the carbonium ion via a combination of 3,2- and 6,2-hydride shifts plus Wagner-Meerwein rearrangement (Scheme 5). The relatively slow 3,2-hydride shift could be isolated ("frozen out" on the NMR time scale) at 203°K, resulting in resolution of the proton NMR spectrum at this temperature into three absorptions (relative areas 4 : 1 : 6).[198,199] In subsequent studies, it was found that the more facile 6,2-hydride shift could be frozen out between 123°K and 145°K.[201] However, efforts to isolate the Wagner-Meerwein rearrangement by low-temperature NMR studies on the norbornyl carbonium ion so far have met with failure. These observations can be interpreted either in terms of extremely facile equilibration of the two classical ions **65a** ⇌ **65b** (Scheme 5) or in terms of a single "nonclassical" (σ-bridged) carbonium ion (represented by either **66a** or **66b**[202]) which can be regarded as a resonance hybrid of **65a** ↔ **65b**.

Scheme 5.

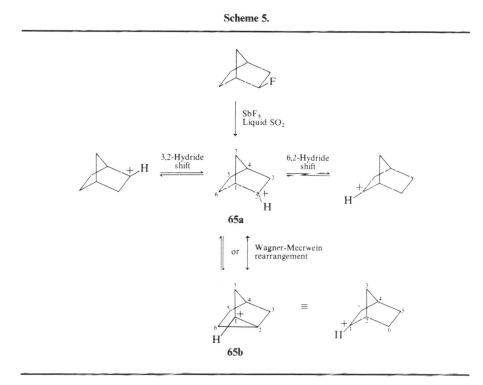

66a   **66b**[202]

In theory, equilibrating classical ions such as **65a** ⇌ **65b** are expected to afford chemical shifts at the charge site which are the "average" of the two contributing structures (one bearing a positive charge at position C-2 and the other being uncharged at this position). Nonclassical delocalization of charge as depicted in **66a** and/or **66b** would be expected to produce an upfield deviation of the observed chemical shift relative to this estimated average. Kirmse has pointed out that "this argument depends critically upon the choice of a model for calculating the average."[196] Herein lies the difficulty in differentiating between classical and nonclassical carbonium ions by NMR spectroscopy. As an example, Olah[198,199] has argued that comparison of the observed and calculated (estimated) average H-1,H-2 proton shifts in **65a** and **65b** supports the nonclassical ion interpretation. However, Kramer[188] has pointed out that the ^1H NMR results obtained for the norbornyl carbonium ion are fully consistent with corresponding observations obtained for the 1,2-dimethylnorbornyl cation **67**, which has been shown[203] to exist as equilibrating classical ions **67a** ⇌ **67b** using the same type of NMR approach as previously employed[200] for the characterization of the norbornyl carbonium ion. Similar confusion results from attempts to interpret the results[203] of low-temperature ^{13}C chemical shift studies on the norbornyl carbonium ion.[196,204]

67a   **67b**

Despite these limitations, much useful information pertaining to charge distribution in bicyclic carbonium ions has been gained through ^{13}C and ^1H chemical shift studies.[205] A number of such studies on bicyclic carbonium ions have been reported; ^{13}C and ^1H chemical shifts in some representative, substituted 2-norbornyl, 7-norbornenyl, 7-norbornadienyl, and 10,11-benzo-2-norbornyl carbonium ions (structures **65–90**) appear in Table 8.

Let us first consider the four 2-substituted-2-norbornyl carbonium ions, **67–70** (Table 8). Comparison of the C-1 and C-2 chemical shifts in ions **68** and **70** with the C-1 shifts in the corresponding 1-substituted 1-cyclopentane carbonium ions (**91** and **92**, see Scheme 6) is instructive.[209] A significantly greater degree of charge delocalization occurs from C-2 to C-1 (presumably via σ bridging[209]) in **68** than is the case in **70**. This can be seen in two ways: first, by taking the difference Δδ between the ^{13}C chemical shift at C-2 in **68**

TABLE 8. ^{1}H AND ^{13}C CHEMICAL SHIFTS IN BICYCLIC CARBONIUM IONS

Carbonium Ion	Solvent System	Temp (°K)	Atom Position	δ^{13}C^a	δ^1H^a	Conclusions	References
65	SbF$_5$–SO$_2$	203	1,2,6	92.8	3.05	Bridged (nonclassical) ion; NMR results suggest a corner-protonated nortri-cyclene structure fot ion **65**.	206–208
		203	1,2	22.			
			6	125.	6.59		
		203	3,5,7	32.1			
		203	4	38.5			
68	FSO$_3$H–SO$_2$ClF	195	1	83.4	4.64	σ-Delocalization, although present, is less pronounced than is the case for ion **65**.	207, 209
			2	273.7			
			3	58.2	2.70		
			4	30.9	2.70		
			5	38.4	1.47		
			6	42.8	3.29, 1.09		
			7	26.2	1.71		
			CH$_3$	45.5	3.00		
67	FSO$_3$H–SO$_2$ClF	195	1,2	170.4	2.42	Comparison of average ^{13}C shifts with those observed or calculated for structural models suggests that **67** is partially σ delocalized. The extent of delocalization in **67** is partially with that of ion **68**.	209–211
			3,7	50.4	2.86		
			4	43.9	1.52		
			5	27.0	2.86		
			6	42.4	2.43		
			CH$_3$	20.4			
69	FSO$_3$H–SO$_2$ClF	195	1	62.4		Ion **69** is a rapidly equilibrating classical carbenium ion undergoing rapid 1,2 Wagner-Meerwein rearrangement. Pi delocalization (into the phenyl rings) dominates over σ delocalization.	209
			2	259.9			
			3	53.6			
			4	42.7			
			5	37.2			
			6	44.4			
			7	23.4			

TABLE 8 (*continued*)

TABLE 8 (continued)

Carbonium Ion	Solvent System	Temp. (°K)	Atom Position	δ^a_{13C}	δ^a_{1H}	Conclusions	References
70 (Ph)	FSO₃H (¹H NMR) SO₃H-SO₂ClF (¹³C NMR)	"Room temp." 238	1	62.4	4.83	Ion **70** is a classical carbonium ion; σ delocalization is essentially absent.	207, 209, 212, 213
			2	259.9			
			3	53.6	3.84 exo		
			4	42.7	3.48 endo		
			5	35.2	3.17		
			6	42.4	3.8 exo		
			7	28.4	2. endo		
71 (H)	FSO₃H-SbF₅ or SbF₅⁻ SO₂ClF	195	1	42.4	3.65	Ion **71** is a classical carbenium ion with charge delocalization occurring into the cyclopropane ring.	214, 215
			2	258.5	11.30		
			3	86.3	4.32		
			4,5	111.6	6.72		
			6,7	46.6	3.44		
72 (CH₃)	FSO₃H-SbF₅ or SbF₅⁻ SO₂ClF	195	1	47.0	3.50	Ion **72** is a classical carbenium ion with charge delocalization occurring into the cyclopropane ring.	214, 215
			2	293.2			
			3	67.5	3.98		
			4,5	83.7	5.60		
			6,7	43.7	3.28		
			CH₃	33.7	3.48		
73 (Ph)	FSO₃H-SbF₅ or SbF₅⁻ SO₂ClF	195	1	38.5	4.10	Ion **73** is a classical carbenium ion with charge delocalization occurring into the cyclopropane ring.	214, 214
			2	275.8			
			3	50.8	3.85		
			4,5	72.8	4.61		
			6,7	44.5	3.01		

Structure	Acid	Temp	Position	^{13}C	^1H	Comment	Ref.
74	FSO₃H–SbF₅ or SbF₅–SO₂ClF	195	1	44.1	3.10	Ion **74** is a classical carbenium ion with charge delocalization occurring into the cyclopropane ring.	214, 215
			2	223.8			
			3	41.3	2.50		
			4,5	38.2	3.85		
			6,7	28.0	2.98		
75	FSO₃H–SbF₅ or SbF₅–SO₂ClF	195	1	40.2	3.34	Ion **75** is a classical carbenium ion with charge delocalization occurring into the cyclopropane ring.	214, 215
			2	258.0			
			3	47.7	3.56		
			4,5	74.0	5.58		
			6,7	42.0	3.59		
76	FSO₃H–SO₂ClF	195	1,4	58.0	4.24	Ion **76** is a nonclassical carbonium ion:	216–218
			2,3	125.9	7.07		
			5,6	26.7	2.44 exo / 1.87 endo		
			7	34.0	3.24		
77	FSO₃H–SO₂ClF	195	1,4	57.1	4.07	Ion **77** is a nonclassical carbonium ion	216–218
			2,3	134.1	7.18		
			5,6	24.8	2.40 exo / 1.82 endo		
			7	71.9			
78	FSO₃H–SO₂ClF	195	1,4	55.0	4.15	Ion **78** is a nonclassical carbonium ion.	216–218
			2,3	132.2	7.52		
			5,6	24.6	2.40		
			7	80.5			

76c

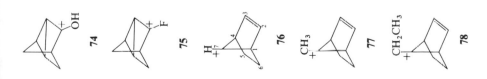

TABLE 8 (*continued*)

TABLE 8 (*continued*)

Carbonium Ion	Solvent System	Temp. (°K)	Atom Position	$\delta_{^{13}C}$	$\delta^a_{^1H}$	Conclusions	References
79	FSO$_3$H–SO$_2$ClF	195	1,4 2,3 5,6 7	51.9 140.1 23.3 109.8	4.72 7.44 2.55 exo 1.97 endo	Ion **79** is a nonclassical carbonium ion	216–218
80	FSO$_3$H–SO$_2$ClF	195	1,4 2,3 5,6 7	46.2 137.9 20.6 225.5	3.65 6.77 2.30 exo 1.72 endo	Ion **80** is a nonclassical carbonium ion.	216–218
81	FSO$_3$H–SO$_2$ClF (¹H NMR) SbF$_5$–SO$_2$–FSO$_3$H (¹³C NMR)	203–223	1,4 2,3 5,6 7	63.8 122.8 115.7 37.0	5.12 7.46 6.10 3.27	Ion **81** is an unsymmetrical (C-2,3 ≠ C-5,6) nonclassical carbonium ion: **81c**	208, 218–220
82	FSO$_3$H–SO$_2$ClF	158	1,4 2,3 5,6 7	69.2 133.6 129.9 74.4		Ion **82** is an unsymmetrical nonclassical carbonium ion.	195

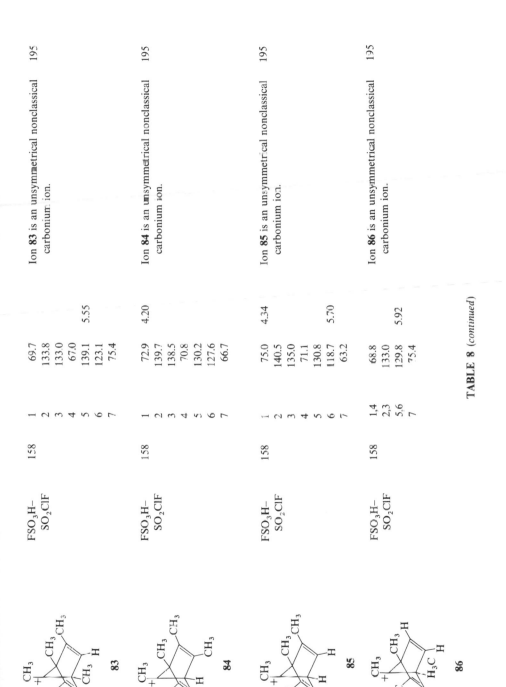

			Position			
FSO₃H–SO₂ClF	158	1	69.7			Ion **83** is an unsymmetrical nonclassical carbonium ion.	195
		2	133.8			
		3	133.0			
		4	67.0			
		5	139.1	5.55		
		6	123.1			
		7	75.4			
FSO₃H–SO₂ClF	158	1	72.9	4.20		Ion **84** is an unsymmetrical nonclassical carbonium ion.	195
		2	139.7			
		3	138.5			
		4	70.8			
		5	130.2			
		6	127.6			
		7	66.7			
FSO₃H–SO₂ClF	158	1	75.0	4.34		Ion **85** is an unsymmetrical nonclassical carbonium ion.	195
		2	140.5			
		3	135.0			
		4	71.1			
		5	130.8			
		6	118.7	5.70		
		7	63.2			
FSO₃H–SO₂ClF	158	1,4	68.8			Ion **86** is an unsymmetrical nonclassical carbonium ion.	195
		2,3	133.0			
		5,6	129.8	5.92		
		7	75.4			

TABLE 8 (*continued*)

TABLE 8 (*continued*)

Carbonium Ion	Solvent System	Temp. (°K)	Atom Position	δ^a_{13C}	δ^a_{1H}	Conclusions	References
87a	FSO₃H–SbF₅ –SO₂ClF or SO₂ClF– SbF₅	195	1,2	96.1		The secondary 2-benzonorbornenyl carbonium ion (**87a**) was not observed; instead, the long-lived ion was found to be benzonortricyclyl carbonium ion (**87b**):	221
			3	51.2			
			4	39.1		**87b**	
			5	131.8			
			6	152.6			
			7	125.2			
			8	165.5			
			9	51.2			
			10	193.8			
			11	84.1			
88a	FSO₃H–SbF₅ –SO₂ClF or SO₂ClF– SbF₅ or FSO₃H– SO₂ClF	195	1	80.8	6.08	Ion **88** is a static, unsymmetrical carbenium ion which possesses less benzonotricyclyl character (**88b**) than does the parent secondary 2-benzonorbornenyl carbonium ion (**87**)[221]	222
			2	199.1			
			3	56.9	3.95 exo 3.20 endo	**88b**	
			4	41.6	3.84		
			5	132.0			
			6	143.1			
			7	124.7			
			8	152.3			
			9	53.4	3.95 syn 3.20 anti		
			10	177.2			
			11	104.5			
			CH₃	26.0	3.08		

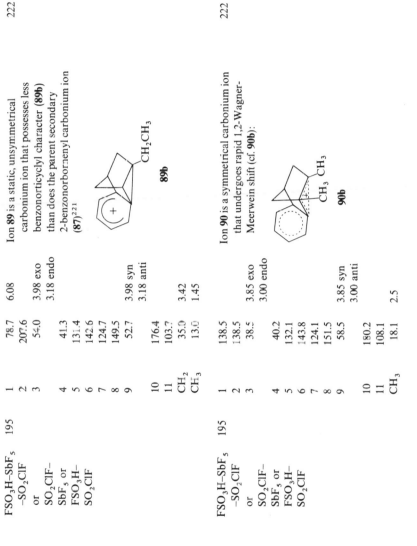

Conditions		Position			Ref
FSO₃H–SbF₅ –SO₂ClF or SO₂ClF– SbF₅ or FSO₃H– SO₂ClF	195	1	78.7	6.08	222
		2	207.6		
		3	54.0	3.98 exo / 3.18 endo	
		4	41.3		
		5	131.4		
		6	142.6		
		7	124.7		
		8	149.5		
		9	52.7	3.98 syn / 3.18 anti	
		10	176.4		
		11	103.7		
		CH₂	35.0	3.42	
		CH₃	13.0	1.45	
FSO₃H–SbF₅ –SO₂ClF or SO₂ClF– SbF₅ or FSO₃H– SO₂ClF	195	1	138.5	3.85 exo / 3.00 endo	222
		2	138.5		
		3	38.5		
		4	40.2		
		5	132.1		
		6	143.8		
		7	124.1		
		8	151.5		
		9	58.5	3.85 syn / 3.00 anti	
		10	180.2		
		11	108.1		
		CH₃	18.1	2.5	

Ion 89 is a static, unsymmetrical carbonium ion that possesses less benzonorticyclyl character (89b) than does the parent secondary 2-benzonorbornenyl carbonium ion (87)[221]

Ion 90 is a symmetrical carbonium ion that undergoes rapid 1,2-Wagner-Meerwein shift (cf. 90b):

TABLE 8 (*continued*)

TABLE 8 (*continued*)

Carbonium Ion	Solvent System	Temp (°K)	Atom Position	δ^a_{13c}	δ^a_{1H}	Conclusions	References
 X = H	FSO$_3$H–SbF$_5$ or SbF$_5$–SO$_2$ClF	195	1 2 3 4 5 6 7	54.3 245.1 46.9 54.3 245.1 46.9 44.0		For all cases of [structure] studied it is concluded that these dications are classical carbonium ions for which σ delocalization is essentially absent.	223
X = OCH$_3$			1 2 3 4 5 6 7	49.3 213.6 42.5 49.3 213.6 42.5 42.1		For all cases of [structure] studied it is concluded that these dications are classical carbonium ions for which σ delocalization is essentially absent.	223

X = CH₃

		223
1	52.2	
2	234.3	
3	45.3	
4	52.2	
5	234.3	
6	45.3	
7	42.8	

For all cases of studied it is concluded that these dications are classical carbonium ions for which σ delocalization is essentially absent.

X = CF₃

FSO₃H–SbF₅   195
or
SbF₅–SO₂ClF

		223
1	56.9	
2	254.4	
3	49.1	
4	56.9	
5	254.4	
6	49.1	
7	44.2	

For all cases of studied it is concluded that these dications are classical carbonium ions for which σ delocalization is essential absent.

FSO₃H–SbF₅   195
or
SbF₅–SO₂ClF

		223
1	58.0	
2	258.2	
3	50.0	
4	58.0	
5	258.2	
6	50.0	
7	45.1	

For all cases of studied it is concluded that these dications are classical carbonium ions for which σ delocalization is essentially absent

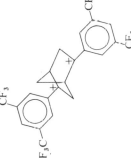

TABLE 8 (*continued*)

TABLE 8 (*continued*)

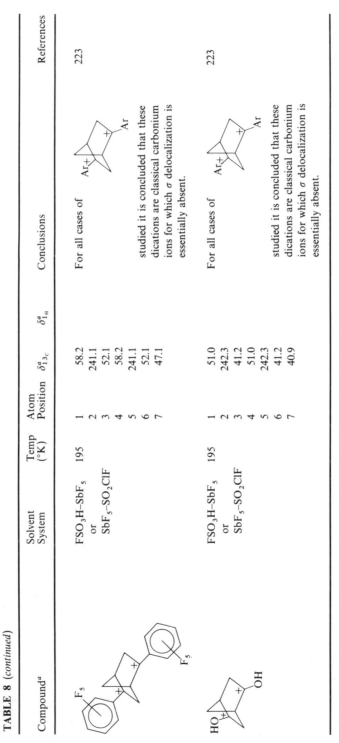

Compound[a]	Solvent System	Temp (°K)	Atom Position	$\delta^a_{13_C}$	$\delta^a_{1_H}$	Conclusions	References
	FSO₃H–SbF₅ or SbF₅–SO₂ClF	195	1	58.2		For all cases of [structure] studied it is concluded that these dications are classical carbonium ions for which σ delocalization is essentially absent.	223
			2	241.1			
			3	52.1			
			4	58.2			
			5	241.1			
			6	52.1			
			7	47.1			
	FSO₃H–SbF₅ or SbF₅–SO₂ClF	195	1	51.0		For all cases of [structure] studied it is concluded that these dications are classical carbonium ions for which σ delocalization is essentially absent.	223
			2	242.3			
			3	41.2			
			4	51.0			
			5	242.3			
			6	41.2			
			7	40.9			

[a] All chemical shift values reported in Table 8 are in ppm downfield from TMS external (capillary) standard.

Scheme 6

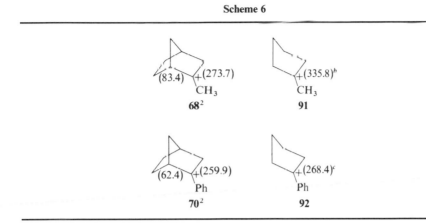

68² 91

70² 92

a ¹³C chemical shift data shown in parentheses (taken from Table 8).
b SbF₅–SO₂ solvent, 213°K; see Table 5 in ref. 199, pp. 72–73.
c FSO₃H solvent, 253°K; see Table 5 in ref. 199, pp. 72–73.

and the corresponding shift at C-1 in the model compound **91** ($\Delta\delta = 335.8 - 273.7 = 62.1$) and comparing the value of $\Delta\delta$ thereby obtained with the corresponding $\Delta\delta$ value for the pair **70** and **92** (for which $\Delta\delta = 268.4 - 259.9 = 8.5$). The larger value in the former instance suggests the relative importance of σ bridging in ion **68** vis-à-vis ion **70**.[209] Secondly, the chemical shift at C-1 in ion **68** is seen to be deshielded by 21 ppm relative to C-1 in ion **70**. Again, this observation has been interpreted[209] to provide further evidence for the relative importance of σ bridging in ion **68** over that in ion **70**.

A similar approach leads to the conclusion[199,209] that σ delocalization is more important in ion **67** than in ion **69**. Olah and Liang[209] have pointed out that the chemical shift of the *para*-phenyl carbon atom in ion **69** is deshielded by 11.4 ppm relative to the corresponding *para*-phenyl carbon atom in ion **70**, attesting to the increased importance of π (over σ) delocalization in the former system.

The 7-norbornenyl (**76**) and 7-norbornadienyl (**81**) carbonium ions are potentially nonclassical by virtue of π stabilization of the cationic center

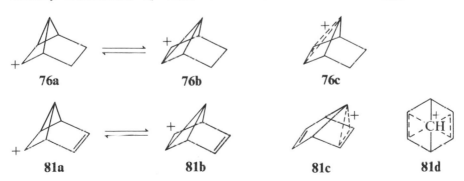

76a 76b 76c

81a 81b 81c 81d

(compare classical ions **76a** $\rightleftharpoons$ **76b** with nonclassical ion **76c**; also, compare classical ions **81a** $\rightleftharpoons$ **81b** with nonclassical ions **81c** and **81d**). Carbon-13 chemical shift and $^1J_{CH}$ coupling constant (see Couplings between Directly Bonded Nuclei, in Chapter 4) data support the nonclassical formulations for ions **76** and **81**. A choice between the unsymmetrical (**81c**) and symmetrical (**81d**) structures for ion **81** is made readily: Both proton[218–220] and ^{13}C[208] NMR studies reveal that the chemical shifts at the four ethene bridge positions are nonequivalent (i.e., H-2,3 $\neq$ H-5,6 and C-2,3 $\neq$ C-5,6), a result that rules out the symmetrical structure (**81d**) for ion **81**.

An interesting relationship between proton chemical shifts and the extent of σ bridging in 2-aryl-2-norbornyl (**93**) and 2-aryl-2-bicyclo[2.2.2]octyl (**94**) carbonium ions has been reported by Farnum and co-workers.[212,213] It was reasoned that the chemical shifts of protons H-1 and H-3 in **93** and **94** should be affected proportionally (i.e., linearly) by the positive charge at C-2 if these ions were classical, whereas the onset of σ bridging in either ion would be expected to change the slope of a plot of δ_{H-1} vs. δ_{H-3} for the affected carbonium ion.[213] Such a plot in the case of ion **93** produces a line with pronounced curvature, whereas the corresponding plot for **94** displays considerably less curvature. In both cases, the deviation from linearity was found to increase when the substituent Y in the aryl group was replaced by groups that were increasingly more electron withdrawing than hydrogen. These results were interpreted to provide evidence for the operation of σ bridging in "those norbornyl cations more electron demanding than the 2-phenylnorbornyl cation."[213]

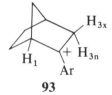

93

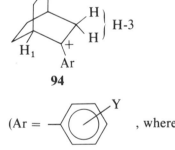

94

(Ar = ⟨phenyl⟩–Y , where

Y = *p*-OCH₃,

     p-OH, *p*-CH₃, H, *p*-F,

     p-Cl, *m*-Cl, *p*-Br,

     m-Br, *p*-I, and *p*-CF₃)

(Ar = ⟨phenyl⟩–Y , where

Y = *p*-OCH₃, *p*-CH₃, H, *p*-F,

     p-Cl, *m*-Cl, *p*-Br, *m*-Br,

     p-I, and *p*-CF₃)

More recently, Farnum has applied the "tool of increasing electron demand"[224,225] to examine the ^{13}C NMR spectrum of meta- and para-substituted 2-arylnorbornyl carbonium ions. Carbon-13 chemical shifts in 2-arylnorbornyl carbonium ions bearing substituents that possess weak electron demand (e.g., *m*- and *p*-halogens, methyl, or methoxy) correlate linearly with one another; a plot of $\delta(^{13}C\text{-}1)$ vs. $\delta(^{13}C\text{-}3)$ in this system bearing this group of substituents in the phenyl ring affords an excellent linear correlation.[226] However, marked deviation from this straight line occurs when the aryl group in the 2-arylnorbornyl carbonium ion bears substituents that are more strongly electron withdrawing (e.g., p-trifluoromethyl and 3,5-bistrifluoromethyl). The point on this plot at which deviation from linearity sets in is considered to reveal the onset of σ bridging in the cation. A similar approach was employed[226] to investigate the onset of π bridging in the 7-arylnorbornenyl carbonium ion.[227–230]

Interestingly, it was noted[226] that in cases where deviations from linearity occur, they do so with less strongly electron-withdrawing substituents for the case of ^{13}C chemical shifts of the cations than they do for the corresponding solvolysis rate constants when the latter are studied in analogous systems. On this basis, Farnum and co-workers concluded that "the onset of π or σ bridging occurs with less electron-demanding aryl groups in the cations than in the transition states presumed to lead to them."[226] Farnum's conclusions are reinforced by the results of recent ^{13}C NMR studies of arylbicyclic carbonium ions emanating from Olah's laboratory.[223,231] Olah and co-workers concluded similarly that the "tool of increasing electron demand" is ineffective for detecting structural changes (e.g., the onset of π and/or σ bridging) via solvolytic studies unless such changes are, in their words, "very significant."[231,232]

In addition to their extensive ^{13}C NMR investigations of 2-arylnorbornyl carbonium ions, Olah and co-workers also have studied a series of aryl-substituted 2,5-diphenyl-2,5-norbornyl dications (**90c–90i**, see Table 8).[223] A plot of $\delta(^{13}C\text{-}1)$ vs. $\delta(^{13}C\text{-}3)$ for **90c–90g** affords an excellent linear correlation, suggesting the absence of σ bridging in these carbonium ions. This result stands in sharp contrast with the behavior of aryl-substituted 2-phenyl-2-norbornyl monocations, which display deviations from linear behavior when the phenyl ring bears electron-withdrawing substituents (because of the onset of nonclassical σ delocalization).[223,226,231] The "classical" nature of these dications "can be rationalized by charge–charge repulsion resulting in increased charge delocalization into the phenyl rings."[223]

A very useful method for assessing potential "nonclassical" behavior in bicyclic carbonium ions has been introduced by Saunders and co-workers.[233–238] Termed "isotopic perturbation of degeneracy," the method utilizes deuterium substitution in methyl substituents (i.e., substituting CD_3 for CH_3) to produce splitting in carbon peaks which would otherwise afford a single resonance signal. For example, in the ^{13}C NMR spectrum of 1-methyl-2-trideuteriomethyl-2-norbornyl carbonium ion, carbon atoms C-1 and C-2 are no longer degenerate, the observed splitting being caused by the isotopic perturbation introduced by the presence of deuterium in the 2-methyl substituent (see Scheme 7). The magnitude of this splitting has been shown*[238] to relate to the extent of σ participation present in the carbonium ion. Hence, a

---

* Recently, a spirited defense of the use of ^{13}C chemical shifts as a criterion for the elucidation of carbocation structures has appeared.[239]

Scheme 7

C-1 and C-2 degenerate

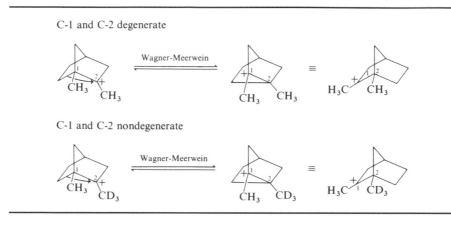

C-1 and C-2 nondegenerate

sensitive probe for detecting σ bridging in bicyclic carbonium ions is afforded by the observation in the ^{13}C NMR spectrum of the effect of this novel secondary deuterium isotope effect upon the carbocation equilibrium established via the 1,2 Wagner-Meerwein shift depicted in Scheme 7.

An important consequence of the foregoing studies is that "classical" and "nonclassical" carbonium ions have come to be regarded as the limits of a continuum of potential carbonium ion structures. A particular carbonium ion, therefore, may be regarded as being more or less "classical" or "nonclassical" depending upon the extent to which σ delocalization contributes to its overall stabilization. The rejection of an "either/or" classical–nonclassical model and its replacement by a "behavioral continuum" concept of carbonium ion stabilization via σ participation would appear to be one of the more pedagogically worthwhile consequences of the NMR shift studies discussed in this section.

The inadequacies attendant upon the present state of understanding of the detailed relationships between proton and ^{13}C chemical shifts and electron densities in bicyclic carbonium ions have been detailed in critical examinations by Farnum,[187] by Kramer,[188] and by Brown and Peters.[204] In addition, the basic concept of applying chemical shift data as criteria for the operation (or absence) of π or σ participation in potentially nonclassical carbonium ions (e.g., the 2-norbornyl, 7-norbornenyl, and 7-norbornadienyl carbonium ions) has been soundly criticized.[204] However, the tone of these critical examinations has been clearly constructive, from which there emerges the implicit hope that future NMR investigators will regard their expressed criticisms more as a challenge than as a deterrant!*[205]

---

* Recently, Williams and Field[240] have reported the existence of linear correlations between ^{13}C chemical shifts in nonclassical carbonium ions and ^{11}B chemical shifts in the isoelectronic and isostructural polyboranes. The use of β-deuterium isotope effects on ^{13}C chemical shifts as "an unambiguous tool for probing the mechanism of charge delocalization in carbocations and a new criterion for differentiating carbocations with different charge-localization mechanisms" has been reported recently.[241] See also refs. 242–244.

4

STEREOCHEMICAL APPLICATIONS OF NMR SPIN–SPIN COUPLING CONSTANTS IN RIGID BICYCLIC SYSTEMS

Introduction

Information gained from NMR spin–spin coupling constants often complements and extends that which can be garnered from studies of NMR chemical shifts. Rigid bicyclic systems related to norbornane and bicyclo[2.2.2]octane have proved invaluable as substrates in defining the stereochemical dependence of coupling constants. Indeed, correlations of spin–spin coupling constants obtained from these studies have permitted stereochemical and configurational assignments to be made in more complex cyclic and acyclic molecules.

Whereas chemical shift values often can be estimated accurately and with relative ease by simple inspection of the relevant NMR spectra, this is true in the case of coupling constants only for the simplest (first-order) spin systems. In higher order spin systems, accurate estimation of coupling constants is often attainable only via computer simulation of the relevant NMR spectra. In many instances, the number of interacting (mutually coupled) spins is sufficiently large as to render computer simulation impractical. In such cases, investigators can resort to any of a number of techniques for spectral simplification; among these are multiple irradiation experiments, isotopic enrichment of a nucleus that possesses a nonzero spin (i.e., $I \neq 0$) and/or replacement of a nonzero spin nucleus by an isotopic nucleus that possesses zero spin. Likewise, as is often the case with investigators whose major interests lie apart from nth-order analysis of complex NMR spectra, coupling constants are "guesstimated" via direct measurement between absorption signals in (often complex) multiplets which appear in the NMR spectra. This last method is the simplest to apply; regrettably, the results that it affords are generally the least satisfactory.

In this chapter, we consider the usefulness of spin–spin coupling constant data in making stereochemical and configurational assignments in substituent-bearing rigid bicyclic systems related to norbornane and bicyclo[2.2.2]octane. Once again, the sheer volume of work published in this area precludes exhaustive treatment of the subject. Wherever possible, estimates of error in cited coupling constants will be reported. Coupling constants whose relative or absolute signs have been either determined experimentally or derived via spectral simulation (computer calculation) are indicated by inclusion of the appropriate algebraic sign in the coupling constant tables that accompany this section. When the absolute value of the coupling constant $|J|$ is given, this indicates that the sign of this coupling constant was not determined in the reference cited. In this section, coupling constants are categorized by the number of bonds intervening between interacting nuclei, [i.e., couplings between directly bonded nuclei (^1J), geminal couplings (^2J), vicinal couplings (^3J), and "long-range" couplings (^4J, ^5J, etc.)].

Couplings between Directly Bonded Nuclei (^1J)

Of the many situations that can be envisioned, involving directly bonded, spin-coupled nuclei couplings X—Y, between directly bonded ^{13}C and ^1H possess the greatest potential for affording insight into structural features of organic compounds. Fortunately, $^1J_{CH}$ couplings are relatively easy to measure; in the days prior to the development of high-resolution Fourier Transform NMR (FT-NMR) instrumentation, this could be accomplished via analysis of natural abundance ^{13}C satellites in proton NMR spectra. Consequently, more data are available for $^1J_{CH}$ than for any other X—Y coupling in this category.*

It was early recognized[247–249] that the magnitude of ^{13}C–^1H spin–spin coupling is determined primarily by the Fermi contact interaction term in the electron–nuclear Hamiltonian.[250] A development of singular importance was the linear correlation of the Fermi contact term with the hybridization (percentage s character) of the carbon nucleus involved in $^1J_{CH}$ coupling.[247–249] More recently, theoretical self-consistent field molecular orbital (SCF-MO) models utilizing finite perturbation techniques have been developed[251–256] that permit refinement of the early theory to take second-order effects into account (e.g., substituent effects on $^1J_{CH}$).

In view of the foregoing, it seems reasonable to expect measurement of $^1J_{CH}$ values in rigid bicyclic systems to be capable of affording at least qualitative (and perhaps even semiquantitative) information regarding relative ring strain effects.[256,257] That this is indeed the case can be seen upon inspection of the

---

* For reviews of spin–spin coupling between directly bonded nuclei see refs. 245 and 246.

data shown in Scheme 8.[206,259-269] Comparison of the methylene bridge $^1J_{CH}$ values for the compounds in Scheme 8 (130–136 Hz) with $^1J_{CH}$ in cyclopentane (128 Hz)[248] affords a measure of the additional strain present in these bicyclic structures (relative to cyclopentane). This increased strain is reflected similarly by comparison of the ethene bridge $^1J_{CH}$ values in norbornene (165 ± 0.5 Hz), norbornadiene (172.5 ± 0.5 Hz), and 2,3-benzonorbornadiene (175.5 ± 0.5 Hz) with the corresponding $^1J_{CH}$ value in cyclopentene ($^1J_{C(sp2)H}$ = 160.5 Hz).[264]

Interestingly, $^1J_{CH}$ values for *syn-* and *anti-*methylene bridge protons in norbornene, 2,3-benzonorbornene, and 2,3-benzonorbornadiene have been observed to be nonequivalent.[166] The smaller of the two bridge carbon–proton couplings occur, respectively, for the bridge carbon–hydrogen bond that is *anti* to the 2,3 double bond in norbornene, for the bridge C—H bond that is anti to the benzene ring in 2,3-benzonorbornene, and for the bridge C—H bond that is syn to the benzene ring in 2,3-benzonorbornadiene (Scheme 8). In each case,

Scheme 8[a]

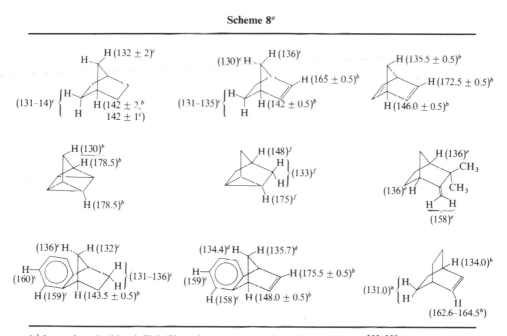

---

[a] $^1J_{CH}$ values in bicyclo[2.2.1]heptyl systems are given in parentheses.[252][258] It has been established that the absolute sign of $^1J_{CH}$ is positive.
[b] Values obtained from ^{13}C satellites in 60 MHz proton NMR spectra.[259-263]
[c] Measured from proton-coupled ^{13}C FT-NMR spectra.[264]
[d] Values obtained from ^{13}C satellites in 100-MHz proton NMR spectra.[264]
[e] Measured from proton-coupled ^{13}C FT-NMR spectra.[265]
[f] Measured[206] at 100 MHz using Olah's internuclear double resonance (INDOR) technique (^{13}C irradiation performed at 25.1 MHz[266]).
[g] Values obtained from ^{13}C satellites in 100-MHz proton NMR spectrum.[267]
[h] Measured from proton-coupled ^{13}C FT-NMR spectra.[268]

it appears that the more sterically hindered of the two bridge carbon–hydrogen bonds in each of these three compounds is the one which displays the smaller $^1J_{CH}$ value. This observation was considered[265] to reflect the consequences of steric crowding upon the electron distribution in the bridge carbon–hydrogen bond and, consequently, upon the hybridization of the bridge carbon atom in these systems. Similarly, it has been observed that $^1J_{CH}$ for the (more sterically crowded) methylene bridge carbon–hydrogen bond in syn-7-chloronorbornene is smaller than is $^1J_{CH}$ for the corresponding carbon–hydrogen bond in the anti isomer; (the $^1J_{CH}$ values are 156.5 ± 1 and 164.2 ± 1 Hz, respectively[265]).

A similar nonequivalence would be expected to exist between $^1J_{CH}$ values for exo and endo (ethano bridge) carbon–hydrogen bonds in norbornene and 2,3-benzonorbornene. Due consideration of steric factors (i.e., the realization that exo carbon–hydrogen bonds are more sterically crowded than are endo carbon–hydrogen bonds) leads to the prediction that $^1J_{CH}(exo)$ should be smaller than $^1J_{CH}(endo)$.[265] However, the complexity of their ^{13}C absorption signals precluded direct determination of exo and endo $^1J_{CH}$ values in these systems.[265]

One-bond carbon–hydrogen couplings in a number of bicyclic carbonium ions have been reported (Scheme 9). In addition to reflecting the hybridization of particular carbon atoms in these systems, the magnitude of $^1J_{CH}$ is considered[198,199] to reflect the distribution of charge in the carbonium ion and, hence, to provide a measure of the extent of nonclassical delocalization in bicyclic carbonium ions. Olah and White[206] have pointed out that the effect of increasing positive charge on a carbon atom in a three-membered ring should result both in (i) deshielding of the ^{13}C resonance and (ii) and increase in $^1J_{CH}$ for that carbon atom.

The early INDOR studies by Olah and White[206] of the low-temperature ^{13}C NMR spectrum of the 2-norbornyl carbonium ion led these investigators to conclude that this species was a nonclassical carbonium ion, better represented by structure **65a** (corner-protonated nortricyclene) than the more conventional[190–197] representation **65b**. The arguments leading to this con-

65a   **65b**

clusion have been summarized by Stothers.[270] From their initial studies, Olah and White were unable to directly obtain $^1J_{CH}$ values from the low-temperature INDOR spectra of the 2-norbornyl carbonium ion; it was necessary for them to estimate $^1J_{CH}$ for the charge-bearing carbon atoms (C-1, C-2, and C-6) in this species. More recent investigations of the ^{13}C NMR spectrum of the 2-norbornyl carbonium ion by Olah and co-workers using the fast Fourier transformation method[201,203] have permitted accurate measurement

Scheme 9^a

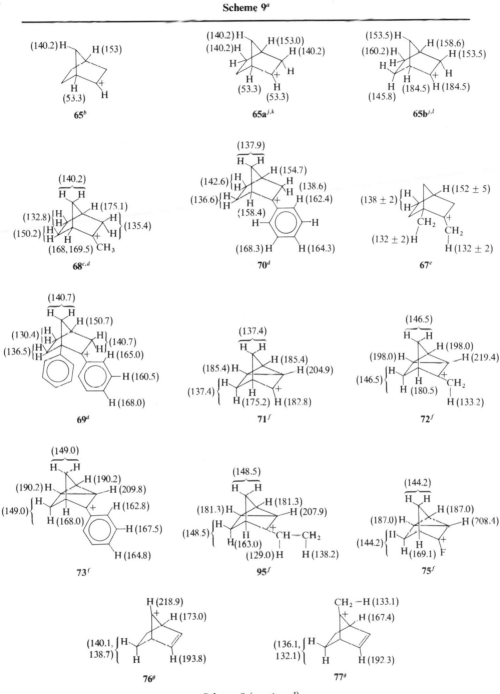

Scheme 9 (*continued*)

Scheme 9 (*continued*)

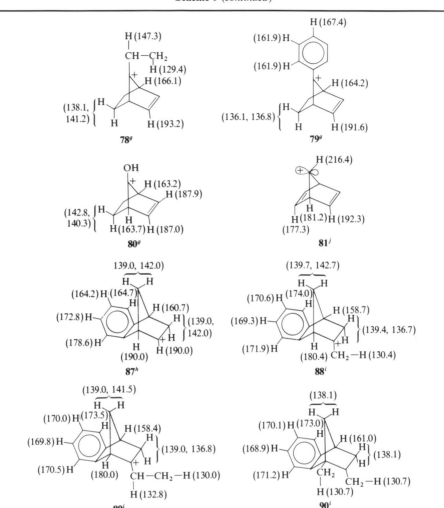

---

[a] $^1J_{CH}$ values in bicyclic carbonium ions are given in parentheses. It has been established that the absolute sign of $^1J_{CH}$ is positive.[252-258]

[b] Measured[206] at 100 MHz using Olah's INDOR technique (^{13}C irradiation performed at 25.1 MHz[266]).

[c] Measured[207] at 100 MHz using Olah's INDOR technique (^{13}C irradiation performed at 25.1 MHz[266]).

[d] Measured from proton-coupled ^{13}C FT-NMR spectra.[209]

[e] Measured[210] at 100 MHz using Olah's INDOR technique (^{13}C irradiation performed at 25.1 MHz[266]).

[f] Measured from proton-coupled ^{13}C FT-NMR spectra.[214]

[g] Measured from proton-coupled ^{13}C FT-NMR spectra.[216,217]

[h] Measured from proton-coupled ^{13}C FT-NMR spectra.[221]

[i] Measured from proton-coupled ^{13}C FT-NMR spectra.[222]

[j] Measured from proton-coupled ^{13}C FT-NMR spectrum.[203]

[k] Measured at 203°K.

[l] Measured at 123°K.

of these $^1J_{CH}$ values to be made directly from its low-temperature (123°K) proton-coupled ^{13}C NMR spectrum (see $^1J_{CH}$ values given for **65a** and **65b** in Scheme 7).

Similar studies of the low-temperature ^{13}C NMR spectra of the 7-norbornenyl and 7-norbornadienyl carbonium ions have led Olah and co-workers[198,199,201,203] to conclude that these species, like the 2-norbornyl carbonium ion, are nonclassical ions (**76a** and **81a**, respectively)[227-229,239] All three species are considered to contain a corner-protonated cyclopropane ring as a common structural feature. Differences in magnitudes among $^1J_{CH}$ values

76a   **81a**

of carbon-hydrogen bonds at the bridging carbon atom in each of these three ions (i.e., at C-6 in **65**, C-7 in **76**, and C-7 in **81**) have been ascribed in part to differences in carbon–hydrogen bond hybridizations owing to inherent variations in ring strain among **65**, **76**, and **81**.* [199]

Arguments based upon relative magnitudes of $^1J_{CH}$ values have seldom been utilized to account for the classical structure of suitably substituted 2-norbornyl carbonium ions (e.g., the 2-phenyl-2-norbornyl[207,209,213,272] and the 1,2-diphenyl-2-norbornyl carbonium ions[209]). Indeed, differential ^{13}C chemical shifts (rather than $^1J_{CH}$ values) generally have provided the basis upon which the suitability of classical vs. nonclassical carbonium ion structural models for specific bicyclic carbonium ions have been judged, (see Correlation of NMR Chemical Shifts with Change Densities in Bicyclic Carbonium Ions, Chapter 3). Attempts to apply $^1J_{CH}$ data as criteria for the operation (or absence) of σ participation in bicyclic carbonium ions have met with much criticism (as was the case with chemical shift data when applied in this way).

In a recent review, Gil and Geraldes[273] have attacked the uncritical manner in which the Muller-Pritchard model[248,249] has been applied quantitatively to relate $^1J_{CH}$ with carbon–hydrogen bond hybridization. Gil and Geraldes note that a number of factors in addition to "percentage s character" influence the magnitude of $^1J_{CH}$. They point out that when comparing $^1J_{CH}$ values along a series of substituted hydrocarbons, one must be aware that the magnitude of $^1J_{CH}$ can be influenced materially by such factors as changes in molecular geometry that accompany substitution and electronic (field, inductive, and hyperconjugative) substituent effects which serve to alter the electronic environment about a given carbon nucleus in the molecule.† [273] As might be

---

* A similar approach based on analysis of ^{13}C NMR chemical shifts and $^1J_{CH}$ couplings has been employed to elucidate the nature of bonding in norbornadienemercurinium ions.[271]

† Detailed analysis of the factors affecting the magnitude of $^1J_{CH}$ is presented by W. McFarlane.[245] An additivity relationship between $^1J_{CH}$ in molecules of the type CHXYZ and substituent constants, ζ, characteristic of the substituents X, Y, and Z has been described by Malinowsky.[274]

expected, the best quality correlations of $^1J_{CH}$ with carbon–hydrogen bond hybridization occur along a series of structurally related hydrocarbons that do not contain highly polar substituents.

In addition to the foregoing, potentially complicating factors, Kramer[188] has objected to the practice of estimating $^1J_{CH}$ values in the 2-norbornyl carbonium ion on the basis of a presumed ground-state nonclassical ion model.[206,272] Kramer has pointed out that "spin coupling arises as a result of a number of complex factors, an important term being derived from transitions between the singlet ground and excited triplet states of the molecule being studied."[188] Because this term could not be taken into account when $^1J_{CH}$ values in "nonclassical" 2-norbornyl carbonium ions were estimated, it seems unlikely that the often cited[198,199] agreement between estimated and observed $^1J_{CH}$ values in this system can possess overriding quantitative significance.

A limited number of studies of $^1J_{CF}$ couplings in rigid bicyclic systems have been reported.[275] In those cases where the sign of $^1J_{CF}$ has been determined, it has been found to be negative.[260,273,276] Some representative $^1J_{CF}$ values in fluorinated norbornanes[80,277] and a 1-fluorobicyclo[2.2.2]octane[165] appear in Table 9. In these cases, the magnitude of observed carbon–fluorine coupling constants has proved useful for assigning resonances in the ^{13}C NMR spectra of fluorinated bicyclic systems, (e.g., for 1-fluorobicyclo[2.2.2]octane, $|^nJ_{CF}|$ values for $n = 1, 2, 3,$ and 4 were found to be 185.3, 18.4, 9.4, and 3.3 Hz, respectively*).[165]

The $^1J_{CF}$ data for the methyl-substituted 2,2-difluoronorbornanes which appear in Table 9 suggest that, at least in some cases, $^1J_{CF}(exo) \neq {}^1J_{CF}(endo)$. The two $^1J_{CF}$ values given for each of these compounds are presented in the order $^1J_{CF}(exo)$ followed by $^1J_{CF}(endo)$. However, the authors note[80] that these assignments are not unequivocal because of the large uncertainties attendant upon their determination. In view of their potential utility as aids to assigning carbon–fluorine bond configurations in fluorinated norbornanes, the need for further investigations designed to clarify the question of the relative magnitudes of $^1J_{CF}(exo)$ and $^1J_{CF}(endo)$ in substituted 2,2-difluoronorbornanes seems apparent.

One-bond carbon–carbon coupling constants ($^1J_{CC}$) have received relatively limited attention. This is a direct consequence of the low natural abundance of ^{13}C, which until recently has mandated the synthesis of specifically ^{13}C-labeled molecules for such studies. Although this requirement certainly has presented a major obstacle to early NMR investigators interested in studying $^nJ_{CC}$ couplings in rigid bicyclic systems, considerable progress has been achieved in recent years both in NMR instrumentation (which permits measurement of $^nJ_{CC}$ in rigid bicyclic systems at ^{13}C natural abundance[80]) and in the synthesis of specifically ^{13}C-labeled norbornanes and related systems.

---

* One potential measure of the effect of ring strain in 1-fluorobicyclo[2.2.2]octane on the magnitudes of J_{CF} values in that system can be gleaned via comparison with the appropriate, corresponding J_{CF} values in 1-fluoroadamantane (a relatively strain-free system) for which $^nJ_{CF}$ values ($n = 1, 2, 3,$ and 4) were found to be 185.9, 17.7, 10.4, and 1.25 Hz, respectively.[278]

TABLE 9. $^1J_{CF}$ Values in Fluorinated Norbornanes and Bicyclo[2.2.2]octanes

Compound	$^1J_{CF}$ (Hz)[a]	Reference
(structure)	|182.0|	80[b]
(structure)		
R = H	|253.9|	80[b]
R = 1-CH$_3$	|256.5|	
R = exo-3-CH$_3$	|258.|	
R = endo-3-CH$_3$	|259.9|	
R = exo-5-CH$_3$	|255.0|	
R = endo-5-CH$_3$	|254.0|	
R = exo-6-CH$_3$	|255.|	
R = endo-6-CH$_3$	|257.3|	
R = syn-7-CH$_3$	|253.|	
R = anti-7-CH$_3$	|254.5|	
(structure)	|185.3|	165
(structure)	|208.1|	277

[a] It has been established that the absolute sign of $^1J_{CF}$ is negative.[260,273,276]
[b] $^1J_{CF}$ values are considered accurate to ± 2 Hz.[80]

The magnitudes of directly bonded carbon–carbon couplings, like $^1J_{CH}$ values, are determined primarily by the Fermi contact contribution. Accordingly, it is not surprising that $^1J_{CC}$ and $^1J_{CH}$ couplings display parallel behavior in similar systems with respect to, e.g., carbon atom hybridization* (and ring strain[80]) and sensitivity to polar substituent (electronegativity) effects.[279,280] The sensitivity of $^1J_{CC}$ to ring strain effects is evident from the data presented in Table 10 for one-bon carbon–carbon couplings in norbornane, nortricyclene, and quadricyclane.[80] The increase in $^1J_{CC}$ for the

* A plot of $^1J_{CC}$ vs. percentage s character afford a linear correlation with a slightly negative intercept.[279] An excellent linear correlation between $^1J_{CH}$ in compounds of the type XYZ—^{13}C—H and $^1J_{CC}$ in compounds of the type XYZ—^{13}C—^{13}CH$_3$ has been reported.[279]

TABLE 10. $^1J_{CC}$ Values in Rigid Bicyclic Ring Systems Related to Norbornane and Bicyclo[2.2.2]octane

Compound[a]	Carbon Positions	$^1J_{CC}$ (Hz)[b]	References
	1,2 1,7	|33.4| |32.5|	80
	1,7 3,4	|40.4| |29.8|	80
	1,2 1,7	|12.6| |41.5|	80
	1,2 1,6	|35.7 or 35.3| |32.35|	281
96	7,8	|59.2|	286
97	2,8	|56.1|	286
98	2,8	|58.5|	286
99	7,8	|40.9 ± 0.1|	287

TABLE 10 (*continued*)

TABLE 10 (*continued*)

Compound[a]	Carbon Positions	$^1J_{CC}$ (Hz)[b]	References
(structure: 100, 101)	2,8	|52.3|	288, 289
	2,8	|52.3|	288, 289

100 X = H_{3x}, Y = CH_3
101 X = CH_3, Y = H_{3n}

Compound	Carbon Positions	$^1J_{CC}$ (Hz)	References
(structure: 102, 103)	2,8	|34.9|	288, 289
	2,8	|34.7|	288, 289

102 X = H_{3x}, Y = CH_3
103 X = CH_3, Y = H_{3n}

(structure with positions 1–8, substituents CH$_3$, X, Y)

	Carbon Positions	$^1J_{CC}$ (Hz)	References
X = $\overset{*}{C}O_2H$, Y = H_{2n}	2,$\overset{*}{C}$(X)	|55.32|	290
X = $\overset{*}{C}H_2OH$, Y = H_{2n}	2,$\overset{*}{C}$(X)	|38.49|	290
X = $\overset{*}{C}HO$, Y = H_{2n}	2,$\overset{*}{C}$(X)	|41.12|	290
X = H_{2x}, Y = $\overset{*}{C}O_2H$	2,$\overset{*}{C}$(Y)	|56.57|	290
X = H_{2x}, Y = $\overset{*}{C}H_2OH$	2,$\overset{*}{C}$(Y)	|39.97|	290
X = H_{2x}, Y = $\overset{*}{C}HO$	2,$\overset{*}{C}$(Y)	|42.19|	290

(structure with positions 1–7, substituents H_3C, CH_3, CH_3, X, Y)

	Carbon Positions	$^1J_{CC}$ (Hz)	References
X = $\overset{*}{C}O_2H$, Y = H_{2n}	2,$\overset{*}{C}$(X)	|56.02|	290
X = $\overset{*}{C}H_2OH$, Y = H_{2n}	2,$\overset{*}{C}$(X)	|38.57|	290
X = H_{2x}, Y = $\overset{*}{C}O_2H$	2,$\overset{*}{C}$(Y)	|57.67|	290
X = H_{2x}, Y = $\overset{*}{C}H_2OH$	2,$\overset{*}{C}$(Y)	|40.33|	290

	Carbon Positions	$^1J_{CC}$ (Hz)	References
	2,9	|74.3|	290

TABLE 10 (*continued*)

TABLE 10 (*continued*)

Compound[a]	Carbon Positions	$^1J_{CC}$ (Hz)[b]	References
H$_3$C CH$_3$ / CH$_3$ / *CH$_2$ (9) / (2)	2,9	|74.30|	290
(8) HC(CH$_3$)$_2$ (7) / O / CH$_3$ (9) (5,6,1,4,2,3)	3,4 4,5 4,7 4.9	|33.0| |31.9| |30.8| |39.7|	291
(8) CH(CH$_3$)$_2$ (7) / CH$_3$ (9) (5,6,1,4,2,3)	3,4 4,7 4,9	|32.4| |32.3| |39.2|	291
(6) CO$_2$Et (5,7,8) / CH$_3$ (9) (1,4,2,3)	3,4 4,5 4,8 4,9	|39.5| |31.8| |31.8| |39.6|	291
(6) CO$_2$Et (5,7,8) / CH$_3$ (9) (1,4,2,3)	3,4	|33.4|	291

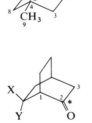

104 X = Cl, Y = H	1,2	|36.5|	292
	2,3	|35.5|	292
105 X = H, Y = Cl	1,2	|39.4|	292
	2,3	|35.6|	292
106 X = Y = H	1,2	|38.6|	292
	2,3	|34.6|	292

TABLE 10 (*continued*)

TABLE 10 (*continued*)

Compound[a]	Carbon Positions	$^1J_{CC}$ (Hz)[b]	References

107 X = H	1,2	|36.0|	292
	2,3	|34.7|	292
108 X = CH$_3$	1,2	|36.7|	292
	2,3	|35.4|	292
	2,9	|40.2|	292

[a] Asterisk indicates position of specific ^{13}C labeling.
[b] It has been established that the absolute sign of $^1J_{CC}$ is positive.[282-284]

1,7 bond along this series reflects the increasing percentage *s* character in the 1,7 bond concomitant with increasing strain (reflected as increasing percentage *p* character, decreasing $^1J_{CC}$ in the 1,2 (3,4) and 1,6 bonds).

A rough correspondence between $^1J_{CC}$ and ring carbon atom hybridization (ring strain) is revealed by the data for the three norbornane ^{13}C-carboxylic acids (**96-98**) presented in Table 10.[286] However, the correspondence is only approximate: The magnitude of $^1J_{CC}$ for the 1,6 bond in cyclopentane ^{13}C-carboxylic acid (56.5 Hz) is smaller than that of the corresponding carbon–carbon bond in **96** and in **98** but slightly larger than the corresponding $^1J_{CC}$ value in **97**.[286] This suggests that additional factors (one of which may be intramolecular nonbonded interactions between the carboxyl group and suitably disposed carbon–hydrogen bonds) may be operating to determine the magnitude of this $^1J_{CC}$ coupling in systems **96-98**. The substituted 2-^{13}C-labeled bicyclo[2.2.2]octanes (**104-108**, Table 10[292]) display remarkably similar values for the (C-1)—(C-2) bond and (C-2)—(C-3) bond $^1J_{CC}$ values.

Only a relatively small number of $^1J_{CN}$ couplings have been measured in rigid bicyclic systems; some representative $^1J_{CN}$ values are presented in Table 11. In general, ^{14}N-^{13}C couplings cannot be directly observed from ^{13}C NMR spectra because rapid quadrupolar relaxation effectively decouples in ^{14}N nucleus.[297,298] However, in molecules where the electric field gradient at the ^{14}N nucleus is small (i.e., in cases where the bonding arrangement about this nucleus is highly symmetrical), it is possible to observe ^{14}N-^{13}C couplings in the ^{13}C NMR spectra.[299,300] Such a situation exists for the tetraalkylammonium iodides **109-111** (Table 11); the $^1J_{CN}$ values cited in Table 11 for these compounds are for ^{14}N-^{13}C couplings. Quadrupolar decoupling is not present for ^{15}N; however, studies involving this isotope of nitrogen are normally performed on specifically ^{15}N-enriched molecules because of the low natural abundance (0.37%) of ^{15}N and its relative insensitivity to detection by NMR methods.

TABLE 11. $^1J_{CN}$ Values in Rigid Bicyclic Ring Systems Related to Norbornane and to Bicyclo[2.2.2]octane[a, b]

Compound[c]	C—N Bond	$^1J_{CN}$ (Hz)[a]	Reference
109	N—C(2)	|1.0|	293
	N—CH$_3$	|3.8|	
110	N—C(2)	|3.2|	293
	N—C(7)	|2.9|	
	N—CH$_3$	|4.3|	
111	N—C(2)	|3.1|	293
	N—CH$_3$	|4.5|	
112	^{15}N—C(2)	|2.1|	294
113	^{15}N—C(2)	|4.8|	294
114	Amide bond (^{15}N—C=O)	|14.3|	295
115	Amide bond (^{15}N—C=O)	|14.0|	295

TABLE 11 (*continued*)

TABLE 11 (continued)

Compound[c]	C—N Bond	$^1J_{CN}$ (Hz)[a]	Reference
116 (bicyclic amide, $C=O$, NH_2, H)	Amide bond (^{15}N—$C=O$)	|13.4|	295
117 (bicyclic amide, $O=C$, NH_2)	Amide bond (^{15}N—$C=O$)	|13.4|	295
(bicyclic lactam, O, HN, positions 2,3,4,5,6,7,8,1)	Amide bond (^{15}N—$C=O$)	|12.1|	296
	^{15}N—C(1)	|7.5|	
(bicyclic, OH, HN_+, Cl^-)	Amide bond (^{15}N—$C=O$)	|16.2|	296
	^{15}N—C(1)	|6.3|	
(bicyclic amine, HN, positions 1,2,3)	^{15}N—C(1)	|2.6|	296
	^{15}N—C(3)	|2.5|	

[a] See text for discussion of the absolute sign of $^1J_{CN}$ couplings.
[b] Unless otherwise indicated, couplings are presented for ^{13}C directly bonded to ^{14}N. An asterisk indicates specific ^{15}N labeling and, in these instances, the couplings given are for ^{13}C directly bonded to ^{15}N.
[c] NMR spectra of tetraalkylammonium iodides were obtained in D_2O solution at 343°K.

When considering the absolute sign of $^1J_{CN}$, it must be borne in mind that ^{15}N and ^{14}N will afford opposite signs of $^1J_{CN}$ simply because their gyromagnetic ratios γ^{15}_N and γ^{14}_N have opposite signs. To circumvent the attendant ambiguity in citing the sign of $^1J_{CN}$ values, a quantity K_{AB}, the "reduced coupling constant," can be employed, where $K_{AB} = (4\pi^2/h\gamma_A\gamma_B) \times J_{AB}$ [i.e., $K_{13C15N} = (-3.268 \times 10^{20}) \times J_{13C15N}$, with K in cm^{-3}, J in Hz].[298,301] Experimental $^1J_{CN}$ values vary over a wide range among different classes of nitrogen compounds (amines, amides, nitro compounds, oximes and imines, nitriles and isonitriles, and unsaturated heterocyclic derivatives).[297,298] The absolute sign of $^1J_{CN}$ generally appears to be negative for most directly

TABLE 12. $^1J_{CP}$ Values in Rigid Bicyclic Ring Systems Related to Norbornane and to Bicyclo[2.2.2]octane

Compound	C—P Bond	$^1J_{CP}$ (Hz)^a	Reference
X = Cl$_2$P	P—C(7)	|47.6|	111
X = Me$_2$P	P—C(7)	|12.2|	
X = Me$_2$(S)P	P—C(7)	|44.6|	
X = Me$_3$P$^+$ I$^-$	P—C(7)	|40.9|	
X = Cl$_2$P	P—C(7)	|40.3|	111
X = Me$_2$P	P—C(7)	| 6.1|	
X = Cl$_2$P	P—C(7)	|45.8|	111
X = Me$_2$P	P—C(7)	|11.|	
	P—C(2)	|63.|	306
	P—C(7)	|64.|	
	P—C(8)	|48.|	307
	P—C(7)	|50.|	
	P—C(2)	|63.|	308, 309
	P—C(2)	|152.6 ± 0.4|	310

TABLE 12 (*continued*)

TABLE 12 (continued)

Compound	C—P Bond	$^1J_{CP}$ (Hz)^a	Reference
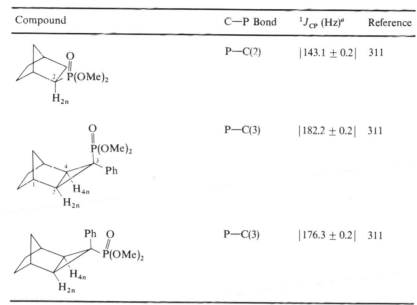	P—C(2)	|143.1 ± 0.2|	311
	P—C(3)	|182.2 ± 0.2|	311
	P—C(3)	|176.3 ± 0.2|	311

^{a} Absolute signs of $^1J_{CP}$: Me$_3$P, negative; Me$_4$P$^+$ I$^-$, positive; (MeO)$_2$P(O)CH$_3$, positive; Me$_3$P(S), positive.[312]

bonded ^{15}N–^{13}C couplings[302-304] (although exceptions to this statement certainly can be found[297,298]).

An early attempt to correlate $^1J_{CN}$ with the product of the percentage s characters of the carbon and nitrogen orbitals that comprise the carbon–nitrogen bond[305] does not appear to have general applicability. A careful theoretical analysis of this correlation has appeared, and criteria for determining the applicability of such $^1J_{CN}$–hybridization correlations to different classes of nitrogen compounds have been suggested.[303]

Some examples of one-bond carbon–phosphorus couplings ($^1J_{CP}$) in rigid bicyclic systems are presented in Table 12. Interestingly, there does not appear to be a consistent relationship between the $^1J_{CP}$ values presented in Table 12 and the anticipated effects of steric crowding. For example, whereas steric crowding has been reported[313-315] to produce a small decrease in $^1J_{CP}$ for alkyl phosphonates, Quin and Littlefield[111] have found the opposite to be true for norbornenes bearing syn- vs. anti-7-trivalent phosphorus substituents [i.e., $^1J_{CP}$ is greater in anti-7-norbornenylphosphorous dichloride and in dimethyl-(anti-7-norbornenyl) phosphine than in the corresponding syn isomers, although the degree of steric crowding in the vicinity of the carbon–phosphorous bond would be expected to be greater in the anti compounds than in the syn series].[111] Thus, Gorenstein's proposal[113] that relates steric congestion effects on $^1J_{CP}$ values to the degree of bond angle distortion concomitant with steric crowding must be viewed critically in the light of Quin and Littlefield's observations[111] cited above.

A number of $^1J_{XH}$ and $^1J_{XC}$ couplings (where $X = {}^{119}Sn$, ^{199}Hg, or $^{203/205}Tl$) in substituted norbornanes and related systems have been reported; some representative examples of these one-bond couplings are presented in Table 13. Couplings ($^1J_{XY}$) on the order of 5–6 kHz are encountered routinely for norbornanes containing thallium–carbon bonds.[320] Directly bonded ^{199}Hg–^{13}C couplings are also quite large in mercurated norbornanes (1.5–1.9 kHz), and they are readily identified in the ^{13}C NMR spectra of these compounds.[317] The observation of these very large $^1J_{XY}$ couplings provides an extremely useful and readily accessible tool for rendering NMR spectral assignments in these systems. Additionally, the potential usefulness of $^1J_{C-Tl}$ and $^1J_{C-Hg}$ couplings as an extremely sensitive probe for measuring carbon atom hybridization in the carbon-to-metal (C—Tl and C—Hg) bond has been suggested.[317,320]

Two-Bond (Geminal) Coupling (^2J)

Geminal proton–proton couplings ($^2J_{HH}$) of the type H—C—H are by far the most extensively studied ^2J couplings in rigid bicyclic systems.[323–325] Representative $^2J_{HH}$ values that have been reported for norbornanes, bicyclo-[2.2.2]octanes, and related systems appear in Table 14. A brief analysis of some of the trends shown by the data in Table 14 follows.

Laszlo and Schleyer[258] have pointed out that, in the absence of complicating substituent electronegativity effects, $|{}^2J_{HH}|$ values should be related to both the H—$\overset{*}{C}$—H bond angle (θ) and the C—$\overset{*}{C}$—C bond angle (ϕ) at the same position ($\overset{*}{C}$ in **120**). That this is indeed the case in norbornene can be

120

seen by comparing $^2J_{7s7a}$ [(C-1)—($\overset{*}{C}$-7)—(C-4) bond angle = ca. 96.5°,[258] $|{}^2J_{7s7a}| = 7.7$–8.8 Hz[76,123–125,264]] with $^2J_{5x5n}$ [(C-4)—($\overset{*}{C}$-5)—(C-6) bond angle ca. 104°,[258] $|{}^2J_{5x5n}| = 10.6$ Hz[63]]. A similar trend has been observed in norcamphor (**118**), which displays four $^2J_{HH}$ couplings (i.e., $^2J_{3x3n}$, $^2J_{5x5n}$, $^2J_{6x,6n}$, and $^2J_{7s7a}$; see Table 14).[342] It can be argued[58] that these bond angle–$^2J_{HH}$ relationships arise from hybridization as well as from geometrical effects.

The effect of electronegative substituents on the magnitudes of $|{}^2J_{HH}|$ values in substituted norbornenes has been extensively studied. Williamson[127] observed that $^2J_{6x6n}$ values in *endo*-5-substituted 1,2,3,4,7,7-hexachloronorborn-2-enes (**56**) could be correlated linearly with the electronegativities of the

TABLE 13. MISCELLANEOUS $^1J_{XY}$ VALUES IN RIGID BICYCLIC RING SYSTEMS RELATED TO NORBOR-NANE AND TO BICYCLO[2.2.2]OCTANE

Compound	X—Y Bond	$^1J_{XY}$ (Hz)	Reference
	$^{119}Sn(2x)$—$^{13}C(2)$	|416.|	316
	$^{119}Sn(2n)$—$^{13}C(2)$	|432.|	316
	$^{199}Hg(2x)$—$^{13}C(2)$	|1660.|	317
	$^{199}Hg(2n)$—$^{13}C(2)$	|1745.|	317
	$^{199}Hg(2x)$—$^{13}C(2)$	|1864.|	317
	$^{199}Hg(3x)$—$^{13}C(3)$	|1564.|	317
R = R' = H, Z = Tl(OAc)$_2$ (solvent. CDCl$_3$)	$^{205}Tl(5x)$—$^{13}C(5)$ $^{203}Tl(5x)$—$^{13}C(5)$	|5645.| |5593.|	318
R = R' = H, Z = Tl(OAc)$_2$ (solvent: pyridine)	$^{205}Tl(5x)$—$^{13}C(5)$ $^{203}Tl(5x)$—$^{13}C(5)$	|6321.| |6362.|	319
R = R' = H, Z = Tl(OAc)$_2$ (solvent: CDCl$_3$)	$^{199}Hg(5x)$—$^{13}C(5)$	|1730.|	319
R = H, R' = OMe, Z = Tl(OAc)$_2$ (solvent: CDCl$_3$)	$^{205}Tl(5x)$—$^{13}C(5)$ $^{203}Tl(5x)$—$^{13}C(5)$	|5945.| |5889.|	318
R = H, R' = Cl, Z = Tl(OAc)$_2$ (solvent: CDCl$_3$)	$^{205}Tl(5x)$—$^{13}C(5)$	|5789.|	318
R = Cl, R' = H, Z = Tl(OAc)$_2$ (solvent: CDCl$_3$)	$^{205}Tl(5x)$—$^{13}C(5)$ $^{203}Tl(5x)$—$^{13}C(5)$	|5847.| |5776.|	318

TABLE 13 (*continued*)

TABLE 13 (*continued*)

Compound	X—Y Bond	$^1J_{XY}$ (Hz)	Reference

R / 4 / L, H_{5n} / 5 / 3 / N, 6 / 1 / 2 / M, O, O

$L = H_{3x}$, $M = H_{5n}$, $N = H_{2x}$, $R = Tl(OAc)_2$	$^{205}Tl(5x)$—$^{13}C(5)$	$\lvert 6631.\rvert$	318
	$^{203}Tl(5x)$—$^{13}C(5)$	$\lvert 6567.\rvert$	
	$^{205}Tl(5x)$—$^{13}C(5)$	$\lvert 6655.\rvert$	320
	$^{203}Tl(5x)$—$^{13}C(5)$	$\lvert 6592.\rvert$	
$L = H_{3x}$, $M = H_{5n}$, $N = H_{2x}$, $R = HgOAc$	$^{199}Hg(5x)$—$^{13}C(5)$	$\lvert 1798.\rvert$	320
$L = H_{3x}$, $M = H_{5n}$, $N = H_{2x}$, $R = HgCl$	$^{199}Hg(5x)$—$^{13}C(5)$	$\lvert 1938.\rvert$	320
$L = H_{3x}$, $M = CO_2Me$, $N = H_{2x}$, $R = Tl(OAc)_2$	$^{205}Tl(5x)$—$^{13}C(5)$	$\lvert 6838.\rvert$	318
	$^{203}Tl(5x)$—$^{13}C(5)$	$\lvert 6775.\rvert$	
$L = H_{3x}$, $M = CO_2Me$, $N = H_{2x}$, $R = Tl(OAc)_2$	$^{205}Tl(5x)$—$^{13}C(5)$	$\lvert 6818.\rvert$	318
	$^{203}Tl(5x)$—$^{13}C(5)$	$\lvert 6753.\rvert$	

H_{2x} / 1 / OAc / H_{3x} / 4 / R / H_{6n} / H_{5n}

$R = HgOAc$	$^{199}Hg(5x)$—$^{13}C(5)$	$\lvert 1732.\rvert$	321
$R = Tl(OAc)_2$	$^{205}Tl(5x)$—$^{13}C(5)$	$\lvert 5812.\rvert$	
	$^{203}Tl(5x)$—$^{13}C(5)$	$\lvert 5759.\rvert$	

H_{2x} / 1 / OAc / H_{3x} / 4 / R / H_{6n} / H_{5n}

$R = HgOAc$	$^{199}Hg(5x)$—$^{13}C(5)$	$\lvert 1720.\rvert$	321

OAc / 1 / R / 4 / H_{6n} / H_{2n} / H_{3n} / H_{5n}

$R = HgOAc$	$^{199}Hg(5x)$—$^{13}C(5)$	$\lvert 1733.\rvert$	321
$R = Tl(OAc)_2$	$^{205}Tl(5x)$—$^{13}C(5)$	$\lvert 5774.\rvert$	

OAc / 1 / R / 4 / H_{6n} / H_{2n} / H_{3n} / H_{5n}

$R = HgOAc$	$^{199}Hg(5x)$—$^{13}C(5)$	$\lvert 1705.\rvert$	321
$R = TlOAc$	$^{205}Tl(5x)$—$^{13}C(5)$	$\lvert 5770.\rvert$	

TABLE 13 (*continued*)

TABLE 13 (*continued*)

Compound	X—Y Bond	$^1J_{XY}$ (Hz)	Reference
R = Tl(OAc)$_2$, R′ = OAc	^{205}Tl(2x)—^{13}C(2)	|5754.|	320
(solvent: CDCl$_3$)	^{203}Tl(2x)—^{13}C(2)	|5701.|	
	^{205}Tl(2x)—^{13}C(2)	|5750.|	319
	^{203}Tl(2x)—^{13}C(2)	|5696.|	
R = HgOAc, R′ = OAc	^{199}Hg(2x)—^{13}C(2)	|1747.|	320
(solvent: CDCl$_3$)			
R = HgOAc, R′ = H$_{3x}$	^{199}Hg(2x)—^{13}C(2)	|1660.|	320
(solvent: CDCl$_3$)			
R = HgCl, R′ = OAc	^{199}Hg(2x)—^{13}C(2)	|1864.|	320
(solvent: Me$_2$S=O)			
R = Tl(OAc)$_2$	^{205}Tl(5x)—^{13}C(5)	|5764.|	320
(solvent: CDCl$_3$)	^{203}Tl(5x)—^{13}C(5)	|5711.|	
	^{205}Tl(5x)—^{13}C(5)	|5471.|	319
R = HgOAc	^{199}Hg(5x)—^{13}C(5)	|1735.|	320
(solvent: CDCl$_3$)			
Z = O	^{199}Hg(1)—^{13}C(1)	|2507.|	322
Z = CH$_2$	^{199}Hg(1)—^{13}C(1)	|2112.|	
Z = O	^{199}Hg(1)—^{13}C(1)	|1068.|	322
Z = CH$_2$	^{199}Hg(1) –^{13}C(1)	|909.|	
	^{77}Se(2)—^{13}C(2)	|68.2|	268

TABLE 14. $^2J_{HH}$ Values in Rigid Bicyclic Ring Systems Related to Norbornane and Bicyclo[2.2.2]octane[a]

Compound	Proton Positions	$^2J_{HH}$ (Hz)	References
	5x,5n (= 6x,6n)	|10.6|	63
	7s,7a	|7.7|	63
	7s,7a	|8.2|	264
	7s,7a	|7.9–8.0|	125
	7s,7a	|8.5 ± 0.3|	123
	7s,7a	|8.|	328
W = H$_{5x}$, X = H$_{6x}$, Y = Z = Br	7s,7a	|9.6|	329
W = H$_{5x}$, X = H$_{6x}$, Y = Cl, Z = Br	7s,7a	|9.4|	329
W = X = Br, Y = H$_{5n}$, Z = H$_{6n}$	7s,7a	|8.9|	329
W = Z = Br, X = H$_{5x}$, Y = H$_{6n}$	7s,7a	|9.7|	329
W = CO$_2$Me, X = CO$_2$H, Y = H$_{5n}$, Z = H$_{6n}$	7s,7a	|9.1|	330
W = H$_{5x}$, X = H$_{6x}$, Y = CO$_2$Me, Z = CO$_2$H	7s,7a	|8.7|	330
W = X = CO$_2$Me, Y = H$_{5n}$, Z = H$_{6n}$	7s,7a	|9.1|	330
W = H$_{5x}$, X = H$_{6x}$, Y = Z = CO$_2$Me	7s,7a	|8.7|	330
W = Z = CO$_2$Me, X = H$_{6x}$, Y = H$_{6n}$	7s,7a	|8.8|	330
W,X = , Y = H$_{5n}$, Z = H$_{6n}$	7s,7a	|10.2|	330
W = H$_{5x}$, X = H$_{6x}$, Y,Z =	7s,7a	|9.0|	330
W = H$_{5x}$, X = OH, Y = H$_{5n}$, Z = CH$_3$	5x,5n	|9.|	332
W = H$_{5x}$, X = CH$_3$, Y = H$_{5n}$, Z = OH	5x,5n	|12.|	332

TABLE 14 (continued)

TABLE 14 (*continued*)

Compound	Proton Positions	$^2J_{HH}$ (Hz)	References
W – H$_{5x}$, X = H$_{6x}$, Y = H$_{5n}$, Z = SiMe$_3$	5x,5n 7s,7a	|10.8| |7.70|	334

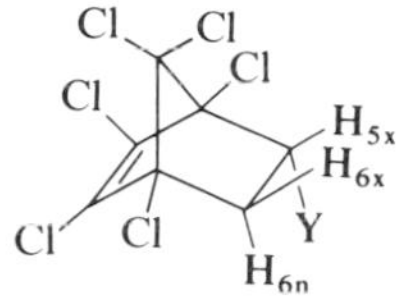

56

Y = H$_{5n}$	5x,5n	−12.4	335, 336
Y = CN	6x,6n	−12.6	127
Y = CO$_2$H	6x,6n	−12.6	127
Y = CO$_2$H	6x,6n	−12.2	336
Y = Ph	6x,6n	−12.7	127
Y = Cl	6x,6n	−13.2	127
Y = Cl	6x,6n	−13.8	337
Y = OH	6x,6n	−12.6	127
Y = OAc	6x,6n	−13.3	127
Y = P(O)(OMe)$_2$	6x,6n	−12.4 ± 0.09	338
Y = CH$_3$	6x,6n	−12.28	357
Y = p-substituted phenyl	6x,6n	−12.7	128, 129

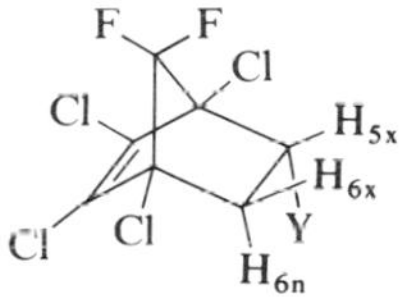			
Y = CN	6x,6n	−13.15	339
Y = CO$_2$H	6x,6n	−13.16	
Y = Ph	6x,6n	−12.43	
Y = Cl	6x,6n	−14.11	
Y = OH	6x,6n	−12.99	
Y = OAc	6x,6n	−13.72	

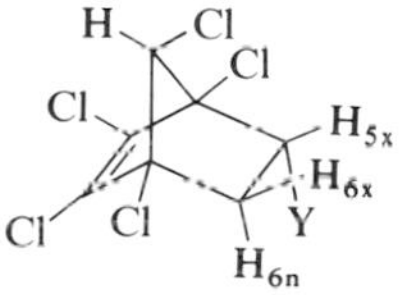			
Y = P(O)(OMe)$_2$	6x,6n	−12.3 ± 0.03	338
Y = CN	6x,6n	−12.59	354
Y = CO$_2$Me	6x,6n	−12.31	
Y = OAc	6x,6n	−13.03	
Y = Br	6x,6n	−13.37	
Y = Cl	6x,6n	−13.18	
Y = Ph	6x,6n	−12.84	

TABLE 14 (*continued*)

TABLE 14 (*continued*)

Compound	Proton Positions	$^2J_{HH}$ (Hz)	References

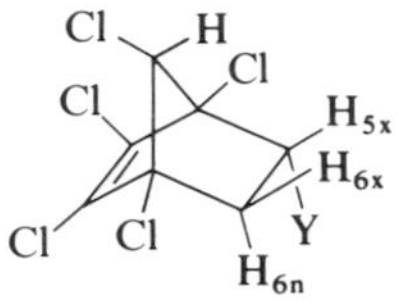			
Y = CN	6x,6n	− 12.71	354
Y = CO$_2$Me	6x,6n	− 12.23	
Y = OAc	6x,6n	− 13.09	
Y = Br	6x,6n	− 13.52	
Y = Cl	6x,6n	− 13.37	
Y = Ph	6x,6n	− 12.80	

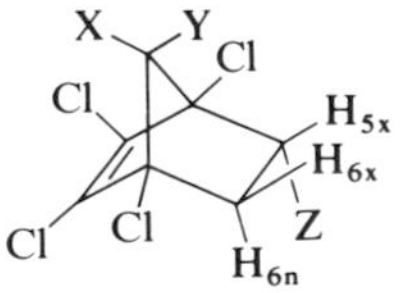			
X = H$_{7s}$, Y = Z = OAc	6x,6n	| 13.5 |	355
X = H$_{7s}$, Y = H$_{7a}$,	6x,6n	| 12.4 |	355
Z = OAc	7s,7a	+ 8.9	

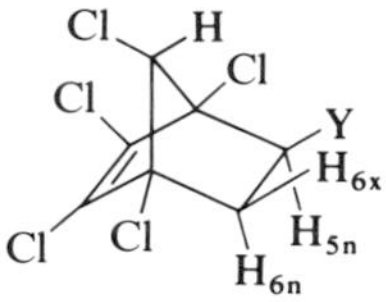			
Y = CN	6x,6n	− 12.53	354
Y = CO$_2$Me	6x,6n	− 12.29	354
Y = OAc	6x,6n	− 13.05	354
Y = OAc	6x,6n	| 12.9 |	355
Y = Br	6x,6n	− 13.40	354
Y = Cl	6x,6n	− 13.39	354

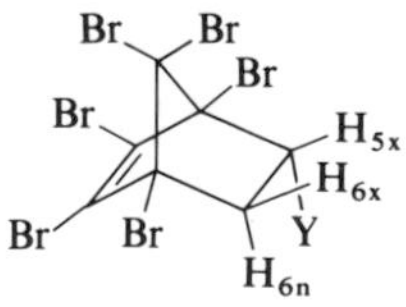

119

Y = H$_{5n}$	6x,6n	− 12.3	340
Y = CN	6x,6n	− 12.5	
Y = CO$_2$Me	6x,6n	− 12.4	

TABLE 14 (*continued*)

TABLE 14 (*continued*)

Compound	Proton Positions	$^2J_{HH}$ (Hz)	References
Y = Ph	6x,6n	−13.0	
Y = OAc	6x,6n	−13.3	

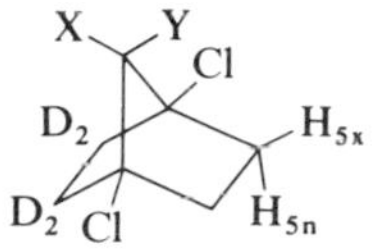 X = H$_{7s}$, Y = H$_{7a}$	5x,5n	−12.09 ± 0.03	341
X = Y = OCH$_3$	5x,5n	−11.78 ± 0.02	
X = OAc, Y = H$_{7a}$	5x,5n	−12.15 ± 0.01	
X = Y = Cl	5x,5n	−12.31 ± 0.04	

| 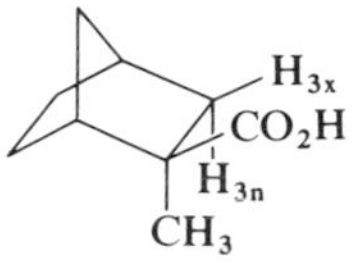 | 3x,3n | \|12.0\| | 118 |

| 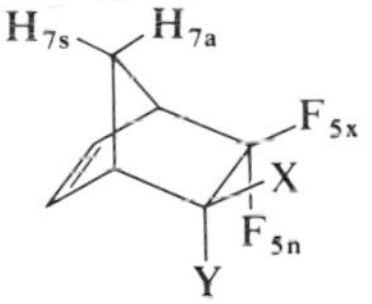 X = H$_{6x}$, Y = F$_{6n}$ | 7s,7a | \|10.5\| | 363 |
| X = F$_{6x}$, Y = H$_{6n}$ | 7s,7a | \|11.\| | |

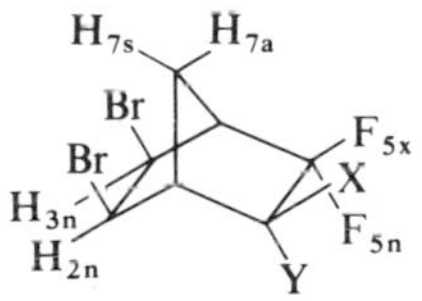 X = F$_{6x}$, Y = F$_{6n}$	7s,7a	|12.5|	363
X = Y = Cl	7s,7a	|13.|	
X = CF$_2$Cl, Y = Cl	7s,7a	|14.|	
X = H$_{6x}$, Y = F$_{6n}$	7s,7a	|12.|	
X = F$_{6x}$, Y = H$_{6n}$	7s,7a	|12.|	

| 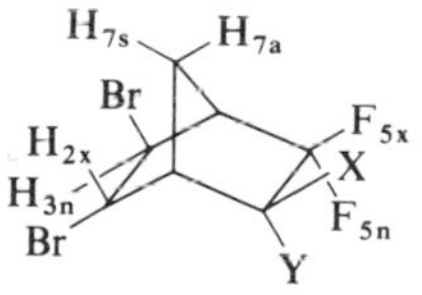 | 7s,7a | \|12.\| | 363 |

TABLE 14 (*continued*)

TABLE 14 (*continued*)

Compound	Proton Positions	$^2J_{HH}$ (Hz)	References
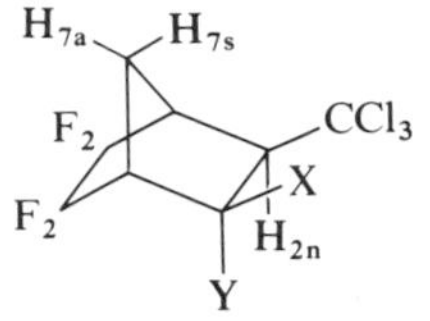 H$_{7a}$ H$_{7s}$ / F$_2$ / F$_2$ / CCl$_3$ / X / H$_{2n}$ / Y			
X = Cl, Y = H$_{3n}$	7s,7a	|12.5|	364
X = Br, Y = H$_{3n}$	7s,7a	|12.5|	
X = H$_{3x}$, Y = Cl	7s,7a	|12.5|	
X = H$_{3x}$, Y = Br	7s,7a	|12.|	
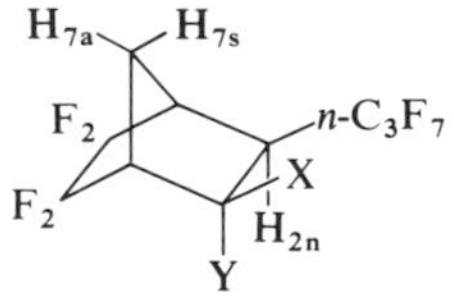 H$_{7a}$ H$_{7s}$ / F$_2$ / F$_2$ / n-C$_3$F$_7$ / X / H$_{2n}$ / Y			
X = I, Y = H$_{3n}$	7s,7a	|12.–13.|	364
X = H$_{3x}$, Y = I	7s,7a	|13.|	
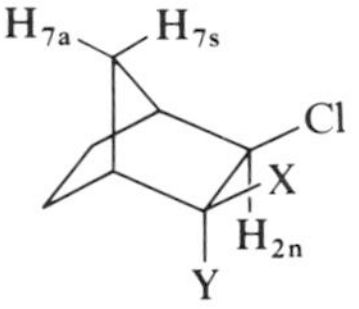 H$_{7a}$ H$_{7s}$ / Cl / X / H$_{2n}$ / Y			
X = Cl, Y = H$_{3n}$	7s,7a	|12.5|	365
X = H$_{3x}$, Y = Cl	7s,7a	|12.5|	
X = Y = Cl	7s,7a	|13.|	
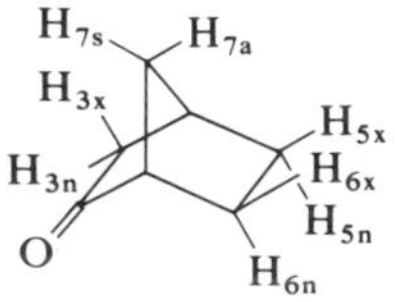 H$_{7s}$ H$_{7a}$ / H$_{3x}$ / H$_{3n}$ / H$_{5x}$ / H$_{6x}$ / H$_{5n}$ / H$_{6n}$ / O **118**	3x,3n	−17.64	342[b]
	5x,5n	−12.79	
	6x,6n	−12.28	
	7s,7a	−10.15	
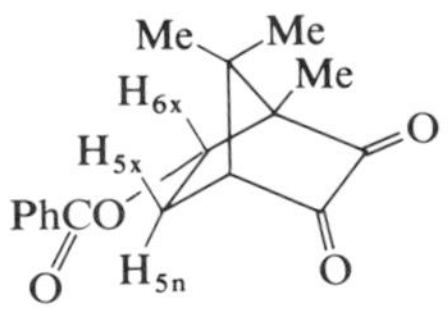 Me Me / Me / H$_{6x}$ / H$_{5x}$ / PhCO / O H$_{5n}$ O O	5x,5n	−15.	359
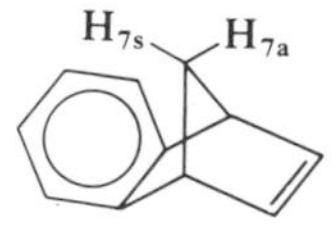 H$_{7s}$ H$_{7a}$			
X = H	7s,7a	|7.2|	124
X = H or Cl	7s,7a	|7.–8.|	362

TABLE 14 (*continued*)

TABLE 14 (*continued*)

Compound 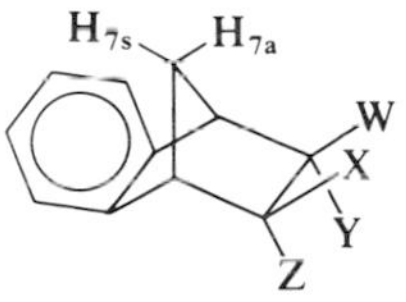	Proton Positions	$^2J_{HH}$ (Hz)	References

$W = H_{5x}, X = H_{6x},$	7s,7a	$	8.7	$	264
$Y = H_{5n}, Z = H_{6n}$	7s,7a	$	8.5 \pm 0.3	$	124
$W - S$ —〈 〉— $CH_3,$	7a,7a	$	10.0	$	360
$X = H_{6x}, Y = H_{5n},$ $Z = Cl$			
$W = CCl_3, X = H_{6x},$ $Y = H_{5n}, Z = Br$	7s,7a	$	10.0	$	
$W = CCl_3, X = H_{6x},$ $Y = H_{5n}, Z = Cl$	7s,7a	$	10.0	$	
$W = SiCl_3, X = D,$ $Y = H_{5n}, Z = H_{6n}$	7s,7a	$	9.0	$	
$W = D_{5x}, X = D_{6x},$ $Y = H_{5n}, Z = H_{6n}$	7s,7a	$	8.5	$	

$R = H_{7s}, S = W = Br,$ $X = H_{6x}, Y = H_{5n},$ $Z = H_{6n}$	6x,6n	$	13.	$	362
$R = H_{7s}, S = Y = Cl,$ $W = H_{5x}, X = H_{6x},$ $Z = H_{6n}$	6x,6n	$	13.	$	
$R = H_{7s}, S = H_{7a},$ $W = H_{5x}, X = Y = Cl,$ $Z = H_{6n}$	7s,7a	$	9.8	$	
$R = H_{7s}, S = H_{7a},$ $W = X = Cl, Y = H_{5n},$ $Z = H_{6n}$	7s,7a	$	10.0	$	
$R = H_{7s}, S = W = Cl,$ $X = H_{6x}, Y = H_{5n},$ $Z = H_{6n}$	6x,6n	$	13.	$	
$R = H_{7s}, S = Cl,$ $W = OAc, X = H_{6x},$ $Y = H_{5n}, Z = H_{6n}$	6x,6n	$	13.0	$	
$R = Cl, S = H_{7a},$ $W = Br, X = H_{6x},$ $Y = H_{5n}, Z = H_{6n}$	6x,6n	$	13.7	$	

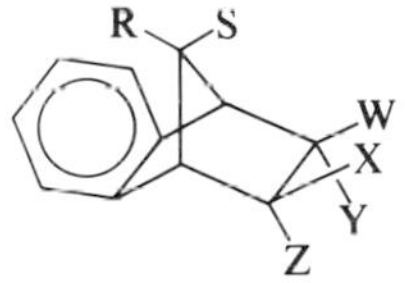

TABLE 14 (*continued*)

TABLE 14 (*continued*)

Compound	Proton Positions	$^2J_{HH}$ (Hz)	References
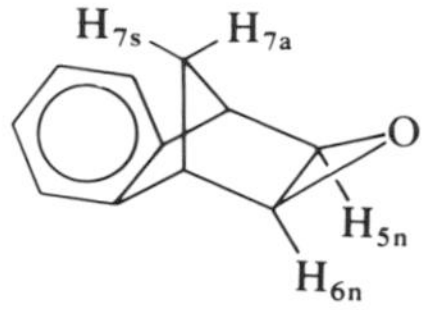 H_{7s} H_{7a} ... O H_{5n} H_{6n}	7s,7a	|8.8|	343
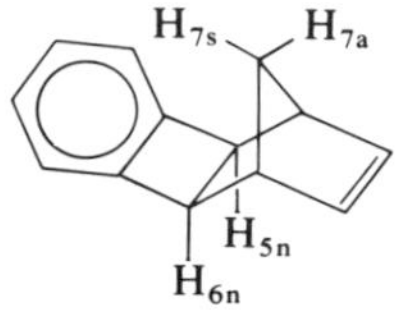 H_{7s} H_{7a} ... H_{5n} H_{6n}	7s,7a	|8.|	344
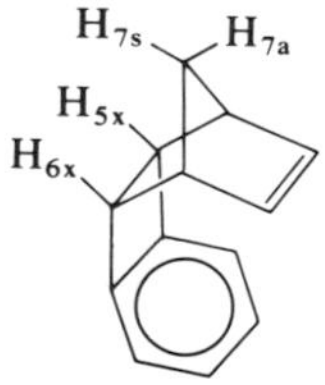 H_{7s} H_{7a} H_{5x} H_{6x} ...	7s,7a	|8.|	344
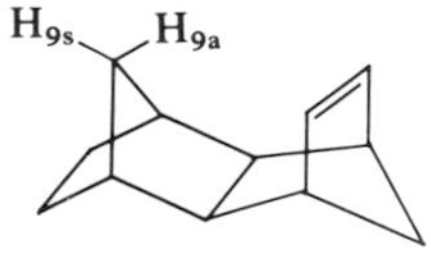 H_{9s} H_{9a} ...	9s,9a	|9.5|	63
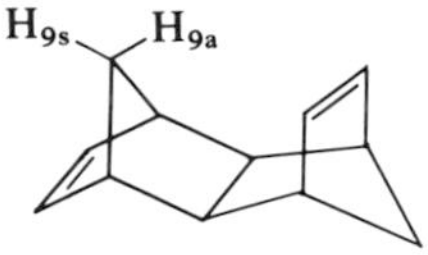 H_{9s} H_{9a} ...	9s,9a	|8.0|	63
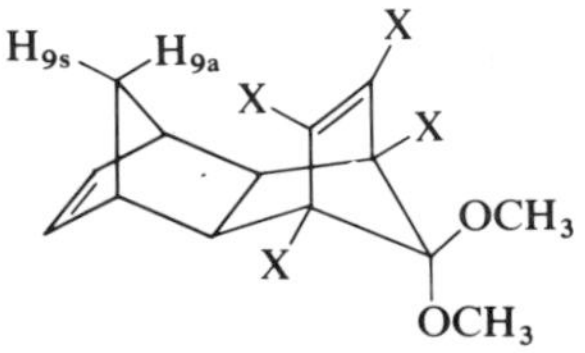 H_{9s} H_{9a} X X X X OCH_3 OCH_3			
X = H	9s,9a	|8.5|	70
X = Cl	9s,9a	|10.|	
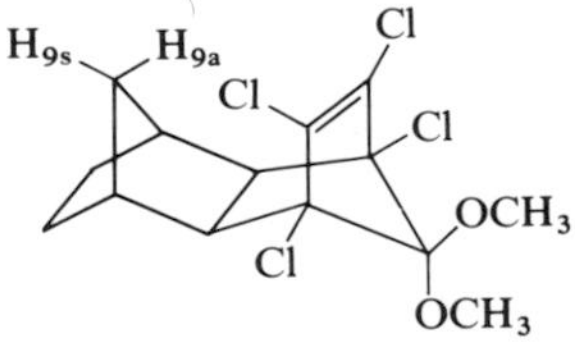 H_{9s} H_{9a} Cl Cl Cl OCH_3 Cl OCH_3	9s,9a	|11.|	70

TABLE 14 (*continued*)

TABLE 14 (*continued*)

Compound	Proton Positions	$^2J_{HH}$ (Hz)	References
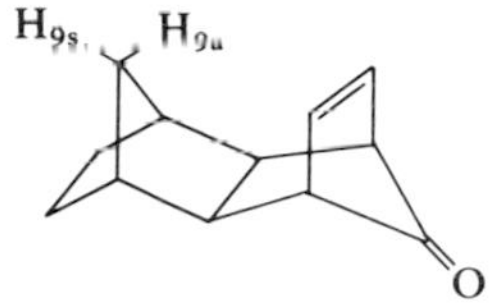	9s,9a	|10.|	70
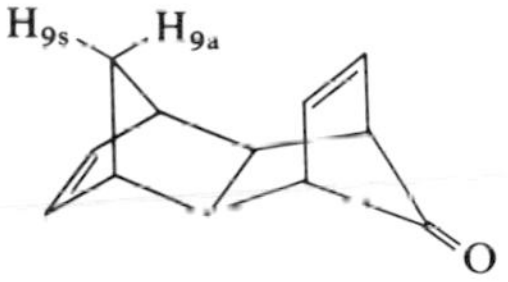	9s,9a	|9.5|	70
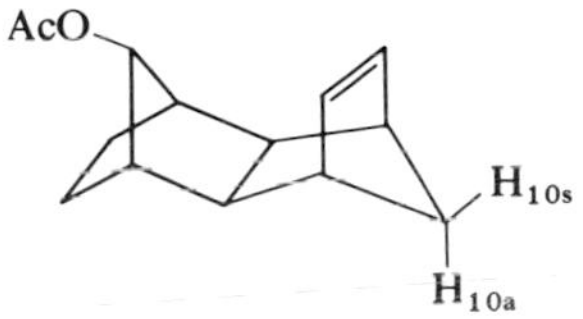	10s,10a	|8.3|	64
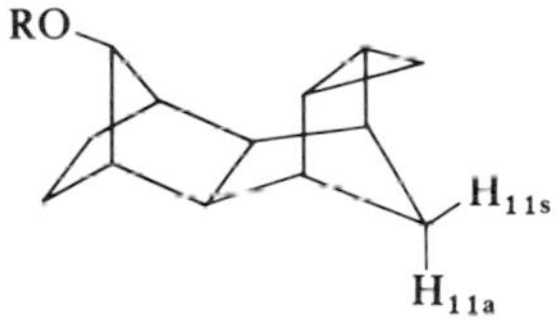 (R = H, CH_3, $C(O)CH_3$, 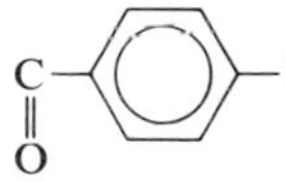 NO_2)	11s,11a	ca. |11.|	64
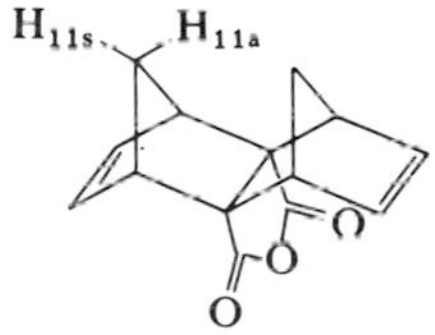	11s,11a	|11.|	68
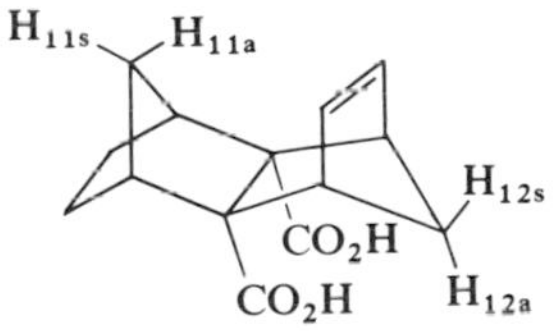	11s,11a 12s,12a	|9.| |15.|	68

TABLE 14 (*continued*)

TABLE 14 (*continued*)

Compound	Proton Positions	$^2J_{HH}$ (Hz)	References
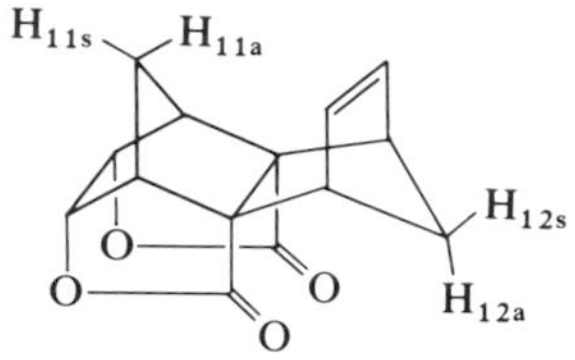	11s,11a 12s,12a	|12.4| |10.|	68
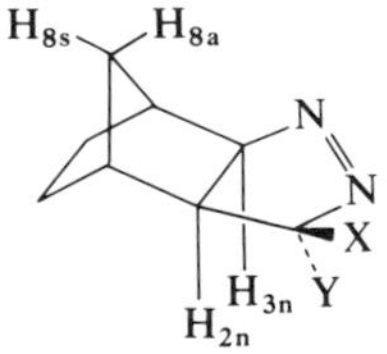			
X = Ph, Y = P(O)(OMe)$_2$	8s,8a	|10.2|	345
X = Me, Y = P(O)(OMe)$_2$	8s,8a	|11.0|	345
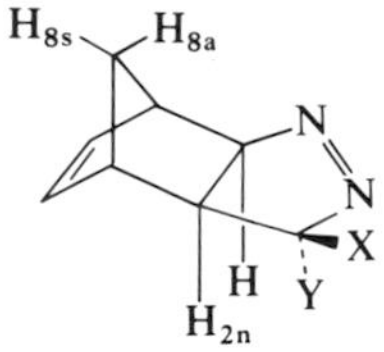			
X = Ph, Y = P(O)(OMe)$_2$	8s,8a	|9.8|	345
X = Me, Y = P(O)(OMe)$_2$	8s,8a	|9.2|	
X = P(O)(OMe)$_2$, Y = Me	8s,8a	|9.3|	
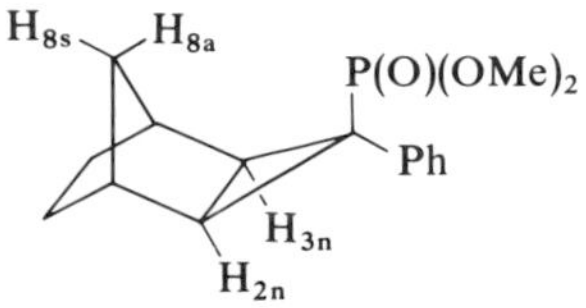	8s,8a	12.2	345
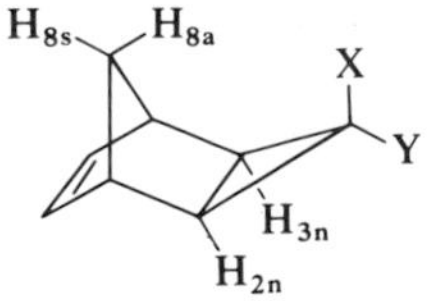			
X = Ph, Y = P(O)(OMe)$_2$	8s,8a	|10.2|	345
X = P(O)(OMe)$_2$, Y = Ph	8s,8a	|10.8|	
X = Me, Y = P(O)(OMe)$_2$	8s,8a	|10.0|	
X = P(O)(OMe)$_2$, Y = Me	8s,8a	|10.6|	

TABLE 14. (*continued*)

TABLE 14 (*continued*)

Compound	Proton Positions	$^2J_{HH}$ (Hz)	References
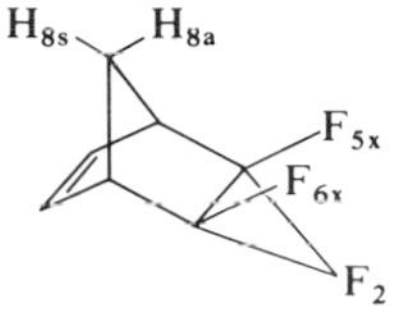 H_{8s} H_{8a} ... F_{5x} F_{6x} F_2	8s,8a	|10.5|	358
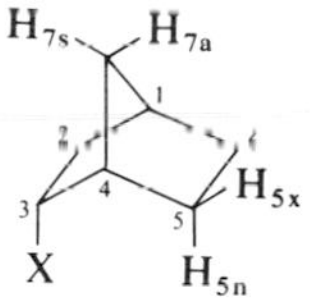 H_{7s} H_{7a} ... H_{5x} X H_{5n}			
X = *endo*-3-octaethyl-porphyrinato-rhodium(I) anion	5x,5n	|10.|	346
	7s,7a	|10.|	
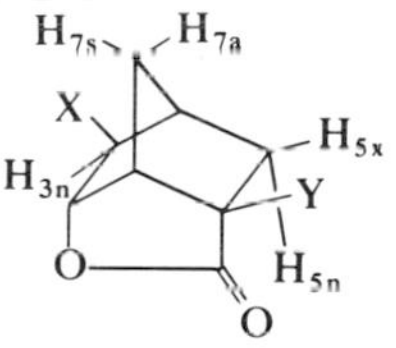 H_{7s} H_{7a} X ... H_{5x} H_{3n} Y H_{5n} O—O			
X = I, Y = H_{6x}	5x,5n	|13.4|	356
	7s,7a	|11.2|	
X = Br, Y = H_{6x}	5x,5n	|13.2|	356
	7s,7a	|11.4|	
X = OAc, Y = H_{6x}	5x,5n	|13.0|	356
	7s,7a	|11.0|	
X = OTs, Y = H_{6x}	5x,5n	|13.4|	356
	7s,7a	|11.0|	
X = I, Y = $(CH_3)_{6x}$	5x,5n	|13.6|	356
	7s,7a	|11.1|	
X = Br, Y = $(CH_3)_{6x}$	5x,5n	|13.6|	356
	7s,7a	|11.1|	
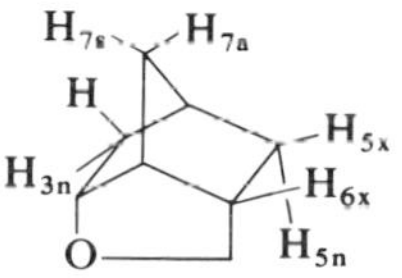 H_{7s} H_{7a} H ... H_{5x} H_{3n} H_{6x} H_{5n} O—			
X = OTs	5x,5n	|13.0|	356
	7s,7a	|10.6|	
X = OAc	5x,5n	|13.2|	356
	7s,7a	|10.6|	

TABLE 14 (*continued*)

TABLE 14 (*continued*)

Compound	Proton Positions	$^2J_{HH}$ (Hz)	References
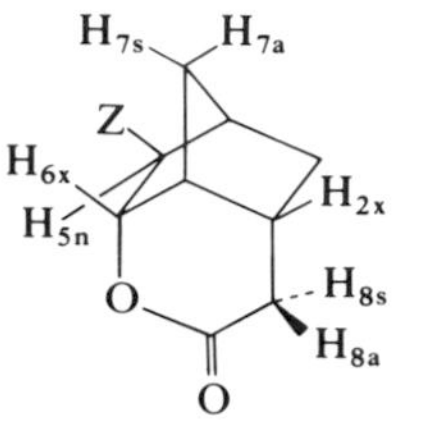			
Z = I	3x,3n	|12.8|	361
	7s,7a	|10.9|	
Z = Br	3x,3n	|11.1|	361
	7s,7a	|9.8|	
Z = HgOAc	3x,3n	|12.8|	361
	7s,7a	|10.7|	
Z = HgBr	3x,3n	|12.5|	361
	7s,7a	|9.5|	
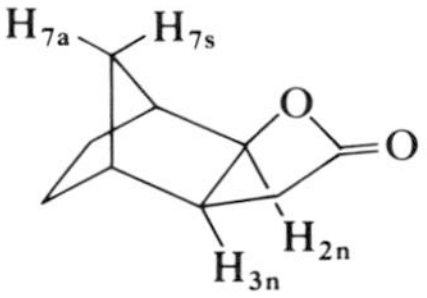	7s,7a	|11.1|	361
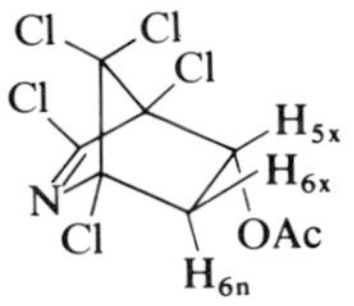	6x,6n	|13.7|	371[c]
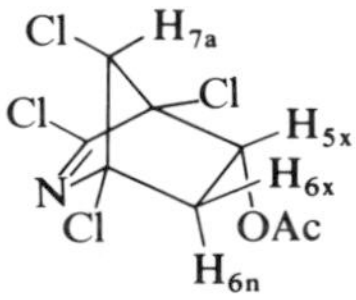	6x,6n	|13.7|	371[c]
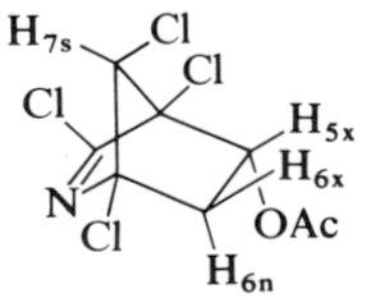	6x,6n	|13.4|	371[c]
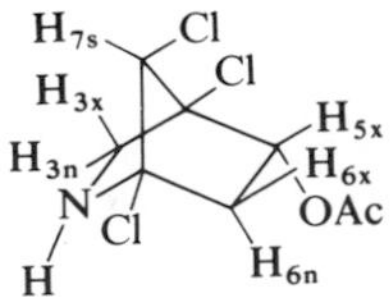	3x,3n	|9.8|	371[c]
	6x,6n	|13.8|	

TABLE 14 (*continued*)

TABLE 14 (continued)

Compound	Proton Positions	$^2J_{HH}$ (Hz)	References
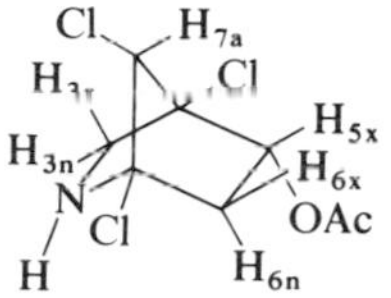 (structure with Cl, H_{7a}, H_{3x}, Cl, H_{5x}, H_{3n}, H_{6x}, N, Cl, OAc, H, H_{6n})	3x,3n 6x,6n	|10.3| |13.9|	371[c]
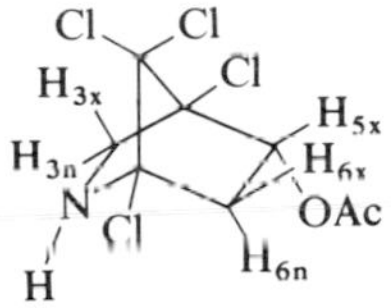 (structure with Cl, Cl, H_{3x}, Cl, H_{5x}, H_{3n}, H_{6x}, N, Cl, OAc, H, H_{6n})	3x,3n 6x,6n	|10.3| |13.9|	371[c]
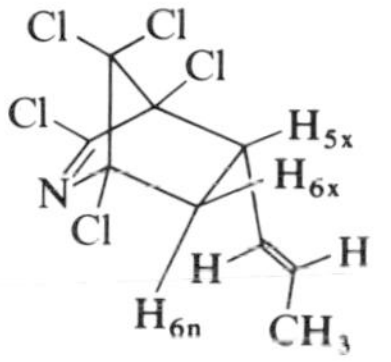 (structure with Cl, Cl, Cl, Cl, H_{5x}, N, H_{6x}, Cl, H, H, H_{6n}, CH_3)	6x,6n	|12.6|	368

(structure with Cl, Cl, Cl, Cl, H_{6x}, H_{5x}, Cl, Cl, X, H_{6n})

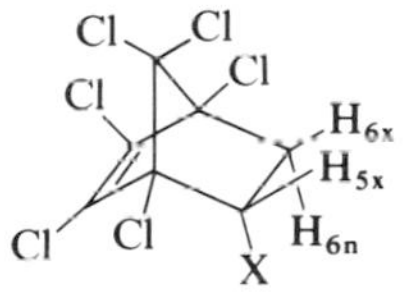

Compound	Proton Positions	$^2J_{HH}$ (Hz)	References
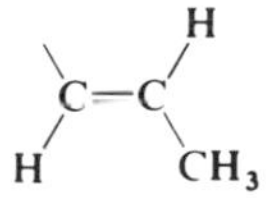 $X = $ (C=C with H, H, H, CH_3)	6x,6n	|12.5|	369
$X = $ 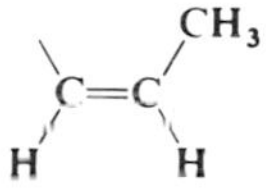 (C=C with CH_3, H, H)	6x,6n	|12.5|	369
$X = $ 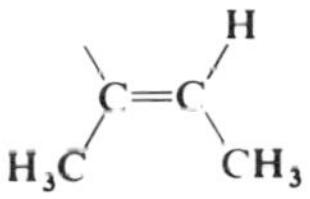 (C=C with H, H_3C, CH_3)	6x,6n	|12.5|	369
$X = $ 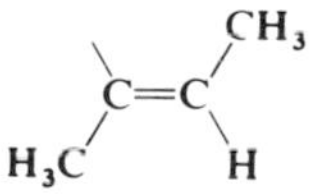 (C=C with CH_3, H_3C, H)	6x,6n	|13.0|	369

TABLE 14 (continued)

TABLE 14 (*continued*)

Compound	Proton Positions	$^2J_{HH}$ (Hz)	References
X 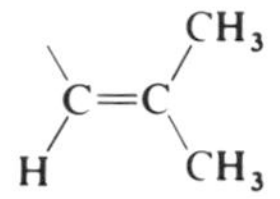 = $\overset{\diagdown}{\underset{H}{C}}=\overset{CH_3}{\underset{CH_3}{C}}$	6x,6n	|12.0|	369
(structure with H_{5x}, H_{6x}, H_{5n}, H_{6n}, N–CH$_3$)	5x,5n	|11.2|	370
(structure with H_{10s}, H_{10a}, H_{11s}, H_{11a})	10s,10a 11s,11a	|10.| |7.5|	347
(structure with H_{10s}, H_{10a}, H_{11s}, H_{11a})	10s,10a 11s,11a	|10.5| |8.0|	347
(structure with H_{10s}, H_{11s}, H_{11a}, H_{10a})	10s,10a 11s,11a	|11.0| |9.5|	347
(structure with H_{11s}, H_{11a}, H_{10s}, H_{10a})	10s,10a 11s,11a	|11.5| |9.5|	347
(structure with H_{11s}, H_{11a}, X, OCH$_3$, OCH$_3$) X = H	11s,11a	|10.|	70
X = Cl	11s,11a	|11.|	

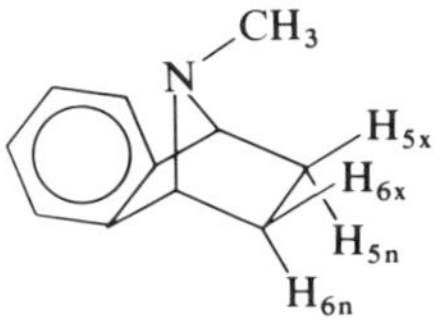

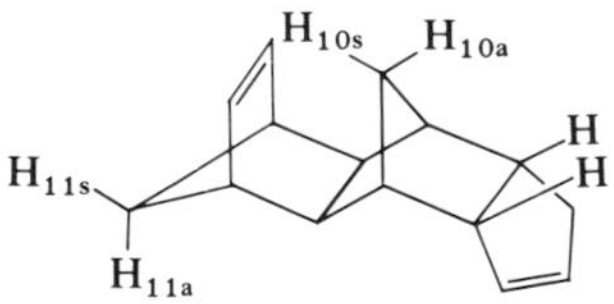

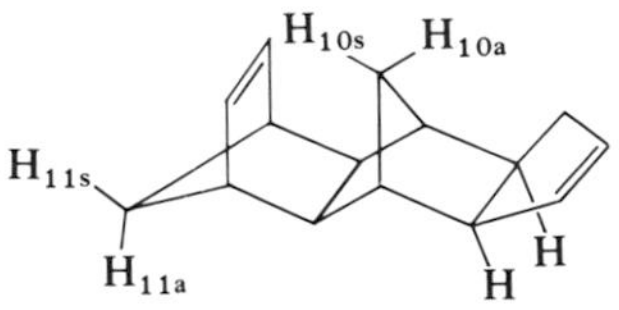

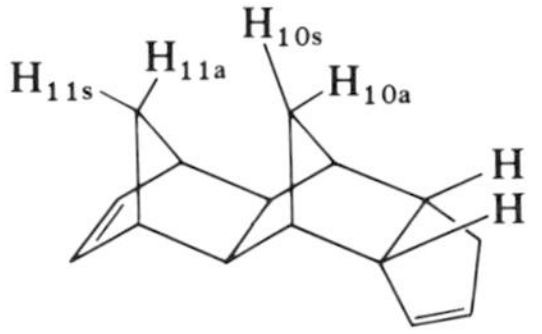

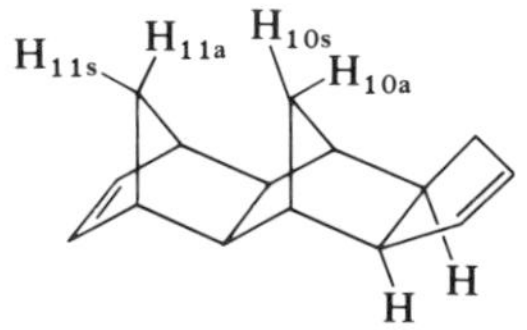

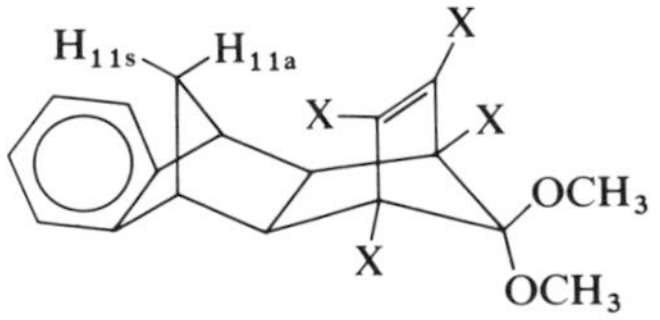

TABLE 14 (*continued*)

TABLE 14 (*continued*)

Compound	Proton Positions	$^2J_{HH}$ (Hz)	References
	11s,11a	|10.|	70
	7s,7a	|8.5|	348
	7s,7a	|9.7|	349
	7s,7a	ca. |9.3|	350
R = H$_{11s}$, S = H$_{11a}$, X = H	11s,11a	|8.|	351[d]
	12s,12a	|12.|	
R = S = X = Cl	12s,12a	|14.|	351[d]
R = H$_{11s}$, S = H$_{11a}$, X = Cl	12s,12a	|12.|	351[d]
	12s,12a	|10.|	351[d]

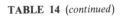

TABLE 14 (*continued*)

TABLE 14 (*continued*)

Compound	Proton Positions	$^2J_{HH}$ (Hz)	References
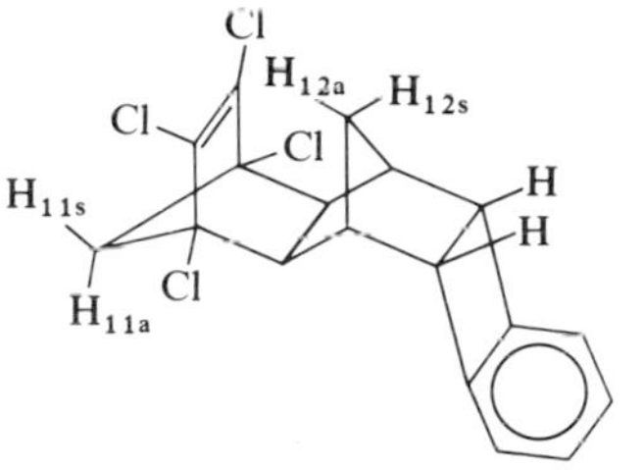	12s,12a	|13.|	351[d]
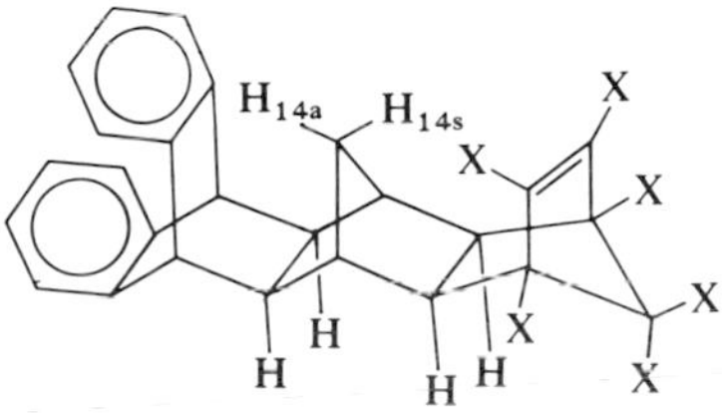			
X = H	14s,14a	|12.|	351[d]
X = Cl	14s,14a	|13.|	
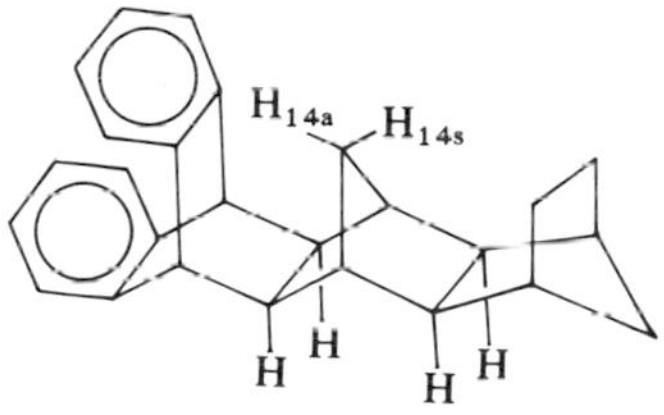	14s,14a	|10.|	351
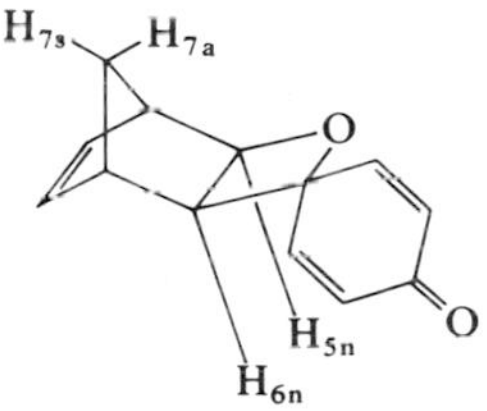	7s,7a	|9.|	366
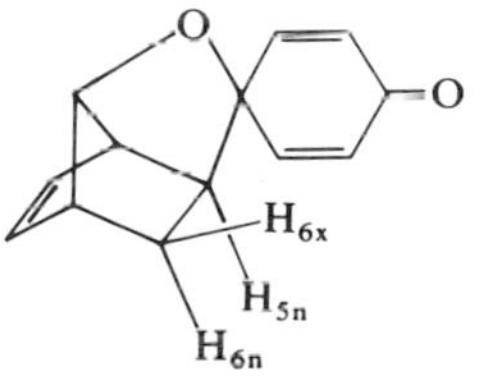	6x,6n	|12.|	366

TABLE 14 (*continued*)

TABLE 14 (*continued*)

Compound	Proton Positions	$^2J_{HH}$ (Hz)	References
X = H, Cl, Br, CN, NO$_2$, CH$_3$, OCH$_3$, NMe$_2$	6x,6n	|10.|	367
X = OH	8s,8a	|13.7|	124
X = OAc	8s,8a	|13.6|	
X = OH	8s,8a	|13.6|	124
X = OAc	8s,8a	|13.8|	
X,Y = *endo* 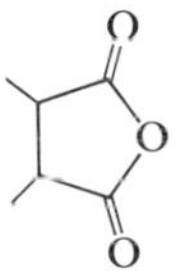	6x,6n	|13.0|	135
X − Y = *endo*-CO$_2$Me	6x,6n	|13.7|	

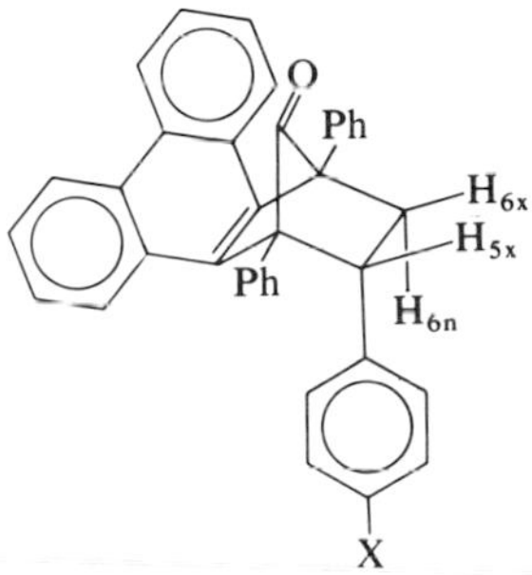

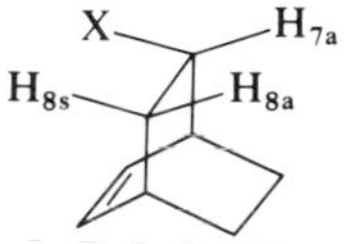

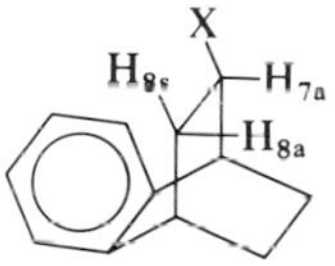

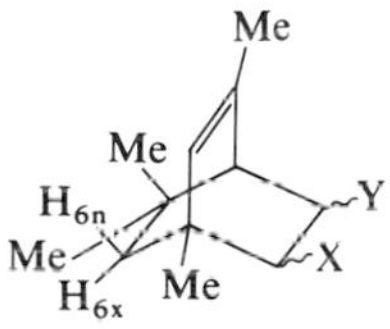

TABLE 14 (*continued*)

TABLE 14 (*continued*)

Compound	Proton Positions	$^2J_{HH}$ (Hz)	References
X = *exo*-CO$_2$Me, Y = *endo*-CO$_2$Me	6x,6n	|12.9|	
X = *endo*-CO$_2$Me, Y = *exo*-CO$_2$Me	6x,6n	|13.0|	
	6x,6n	|17.|	352
	6s,6a	|13.5|	135
	6s,6a	|13.7|	135
X = NH$_2$	8s,8a	|12.3|	353
X = OH	8s,8a	|12.8|	
X = Cl	8s,8a	|14.0|	
X = OAc	8s,8a	|13.0|	
X = OTs	8s,8a	|13.8|	
X = SPh	8s,8a	|12.6|	
X = NMe$_3$	8s,8a	-14.2	224

[a] The results of a number of double-resonance experiments suggest that the absolute value of $^2J_{HH}$ across an sp^3-hybridized carbon is probably negative.[119,127,326,327]

[b] See ref. 342 for estimates of probable errors in these $^2J_{HH}$ values.

[c] Jung and Shapiro[371] have reported that "a W-arrangement of a proton with a chlorine atom causes a downfield shift of approximately 0.15–0.20 ppm for the proton vs. the same proton in the analogous compound without this W-arrangement." The origin of this effect was ascribed to overlap of the rear lobes of the carbon atoms which comprise the C—X and C—H bonds in the W arrangement.

[d] These compounds also display marked (steric) deshielding of H$_{12a}$ relative to H$_{12s}$ (see Chapter 3, Steric Influences of Substituents on NMR Chemical Shifts, Proton Chemical Shifts).[320]

endo-5 substituent groups. A similar linear correlation between $^2J_{6x6n}$ and substituent group electronegativity was achieved for *endo*-5-substituted 1,2,3,-4,7,7-hexabromonorborn-2-enes (**119**, Table 14).[340] In both cases, the range of observed $^2J_{HH}$ values was small (ca. -12.4 to -13.3 Hz). However, the variation of both $^2J_{HH}$ and $^3J_{HH}$ values in these systems was considered to be too large to be rationalized solely in terms of substituent-induced torsional deformation of the relatively rigid norbornene ring skeleton. A similar linear correlation of substituent electronegativity with $^2J_{HH}$ also has been observed for a series of substituted three-membered ring compounds.[372]

In the three cases cited above, increasing substituent electronegativity was observed to have the effect of increasing the absolute magnitude of geminal proton–proton coupling constants $|^2J_{HH}|$ across an sp^3-hybridized carbon atom. This $^2J_{HH}$-increasing electronegativity effect is also apparent in the norcamphor system (**118**): compare $^2J_{3x3n}$ (i.e., the geminal proton–proton coupling constant at the methylene carbon atom that is situated adjacent to the carbonyl group in norcamphor) with the more distant $^2J_{5x5n}$ and $^2J_{6x6n}$ couplings in this system (-17.64, -12.79, and -12.28 Hz, respectively).[342]

Steric compression effects on the magnitude of $^2J_{HH}$ are well documented by the data presented in Table 14. The often sizeable deshielding effects of steric compression upon affected proton chemical shifts in rigid bicyclic systems were noted and discussed in Chapter 3 (Steric Influences of Substituents on NMR Chemical Shifts, Proton Chemical Shifts). Concomitant with steric deshielding effects on proton chemical shifts is an increase in the value of $|^2J_{HH}|$ for the affected protons. Cursory inspection of the data in Table 14 reveals that $|^2J_{HH}|$ for the bridge (7s,7a) protons in simple norbornenes is on the order of 8 Hz;[37] this value increases to 10–14 Hz in sterically crowded situations.[63,68,70,347,349–351]

As was seen to be the case with steric effects on proton chemical shifts, quantitative (or even semiquantitative) evaluation of the effects of steric compression on $^2J_{HH}$ is far from straightforward. Qualitatively, it can be argued[58] that the observed increase in $^2J_{HH}$ is a manifestation of steric compression effects upon carbon–hydrogen bond hybridization, upon the H—C—H bond angle, or upon the distribution of electrons in the carbon–hydrogen bond (or upon some combination of these admittedly interrelated phenomena).

Relatively few $^2J_{CH}$ values have been reported for carbon–hydrogen couplings measured in rigid bicyclic systems; some representative values for $^2J_{CH}$ couplings that have been reported in norbornane and related systems are presented in Table 15. Marshall and Seiwell[336] have reported that a good quality linear correlation (correlation coefficient $= 0.975$) was obtained when $^nJ_{CH}$ ($n = 2$, 3, or 4) in a series of ^{13}C-labeled carboxylic acids was plotted against the corresponding $^nJ_{HH}$ values in analogous (geometrically equivalent) model systems. From this observation, they concluded that "carbon–proton and proton–proton couplings operate by similar mechanisms" throughout the series of compounds chosen for study.[336] Small deviations from the general correlation were noted for geminal olefinic $^2J_{CH}$ values in this series.[336]

TABLE 15. $^2J_{CH}$ Values in Rigid Bicyclic Ring Systems Related to Norbornane and Bicyclo[2.2.2]octane

Compound	Carbon Atom Position	Proton Position	$^2J_{CH}$ (Hz)	Reference
X = $^{13}CO_2H$	$^{13}C{=}O$	5x	-6.35^a	336
X = Cl	5	6x	-6.1^b	337
	5	6n	$< \lvert 2.0 \rvert$	337
H_7—$^{13}CO_2H$	$^{13}C{=}O$	7	-4.61^a	336
	$^{13}C{=}O$	2x	-6.89^a	375
	7(C=O)	1	$\lvert 2.7 \rvert$	375
	7(C=O)	1	$\lvert 2.8 \rvert$	375
	7(C=O)	1	$\lvert 2.8 \rvert$	375
	7(C=O)	4	$\lvert 2.8 \rvert$	375

^{a} The absolute sign of $^2J_{CH}$ across a carbonyl group (i.e., $^2J_{H-C-C=O}$) has been shown to be negative.[373,374]
^{b} The relative sign of this $^2J_{CH}$ coupling was shown to be opposite of $^3J_{H(5x)H(6x)}$.[337]

In addition to proton–proton and carbon–proton couplings, a number of additional two-bond (geminal) couplings in norbornyl and bicyclo[2.2.2]octyl systems have been reported. These include $^2J_{FH}$ (Table 16), $^2J_{PH}$ (Table 17),[345] $^2J_{CC}$ (Table 18),[287] $^2J_{FC}$ (Table 19),[165] $^2J_{NC}$ (Table 20), $^2J_{PC}$ (Table 21), $^2J_{FF}$ (Table 22), and some miscellaneous $^2J_{XY}$ couplings (Table 23).

TABLE 16. $^2J_{FH}$ VALUES IN RIGID BICYCLIC RING SYSTEMS RELATED TO NORBORNANE AND BICYCLO[2.2.2]OCTANE

Compound	Fluorine Atom Position	Proton Position	$^2J_{FH}$ (Hz)	Reference
X = H$_{6x}$, Y = H$_{6n}$	5n	5x	+55.1	376
	5n	5x	|54.31|	377
X = F$_{6x}$, Y = F$_{6n}$	5n	5x	|51.72|	377
X = F$_{6x}$, Y = H$_{6n}$	5n	5x	|51.73|	377
	6x	6n	|52.06|	377
	5x	5n	|48.|	378
	6x	6n	|48.|	378
	5n	5x	|52.|	378
	6n	6x	|54.|	378
X = H$_{6x}$, Y = F$_{6n}$	6n	6x	|55.0|	363
X = F$_{6x}$, Y = H$_{6n}$	6x	6n	|55.1|	363
X = H$_{6x}$, Y = F$_{6n}$	6n	6x	|53.|	363
X = F$_{6x}$, Y = H$_{6n}$	6x	6n	|51.|	363
	6x	6n	|51.|	363

TABLE 17. $^2J_{PH}$ Values in Rigid Bicyclic Ring Systems Related to Norbornane and Bicyclo[2.2.2]octane

Compound	Phosphorus Atom Position	Proton Position	$^2J_{PH}$ (Hz)^a	Reference
X = Y = Z = H	5n	5x	|15.0|	379
X = Y = Z = Cl	5n	5x	-16.7 ± 0.14	338
	5n	5x	-16.9	380
X = Cl, Y = Z = OMe	5n	5x	|13.0|	379
X = Z = Cl, Y = H	5n	5x	-15.6 ± 0.05	338
	4	3	|8.9|	381
	1	7	|9.4|	306

^{a} Absolute signs of $^2J_{PH}$: Me$_3$P, positive; Me$_3$P=S, negative; Me$_4$P$^+$ I$^-$, negative; (MeO)$_2$P(O)CH$_3$, negative.[312,382]

TABLE 18. $^2J_{CC}$ Values in Rigid Bicyclic Ring Systems Related to Norbornane and Bicyclo[2.2.2]octane

Compound^{a,b}	Carbon Atom Positions	$^2J_{CC}$ (Hz)	Reference
96	1,8	ca. 0	286
97	1,8	|1.0|	286
	3,8	|1.8|	
98	1,8	ca. |0.4|	286
	3,8	ca. 0	
99	1,8	$<$|0.5 $\pm$ 0.1|	286
	1,8	$<$|0.5 $\pm$ 0.1|	286
	5,8	$<$|0.5 $\pm$ 0.1|	
100	1,8	|1.2|	288, 289
	3,8	$<$|0.6|	

TABLE 18 (*continued*)

TABLE 18 (*continued*)

Compound[a,b]	Carbon Atom Positions	$^2J_{CC}$ (Hz)	Reference
101	1,8	|0.6|	288, 289
	3,8	< |0.6|	
102	1,8	|0.5|	288, 289
	3,8	< |0.2|	
103	1,8	|0.7|	288, 289
	3,8	< |0.3|	
104	2,4	|1.2|	292
	2,6	|1.1|	
	2,7	|2.0|	
105	2,4	|1.6|	292
	2,6	|2.7|	
	2,7	|1.8|	
106	2,4	|1.5|	292
	2,6	|1.7|	
107	2,4	|1.0|	292
	2,6	|1.9|	
	2,7	|0.7|	
108	2,4	|1.2|	292
	2,6	|1.7|	

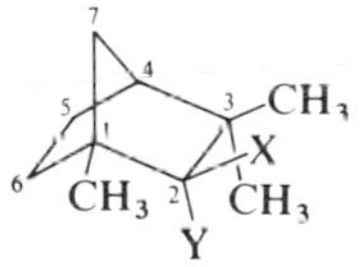			
$X = \overset{*}{C}O_2H,\ Y = H_{2n}$	$1,\overset{*}{C}(X)$	|0.44|	290
	$3,\overset{*}{C}(X)$	|1.86|	
$X = \overset{*}{C}H_2OH,\ Y = H_{2n}$	$1,\overset{*}{C}(X)$	|0.38|	290
	$3,\overset{*}{C}(X)$	|1.10|	
$X = \overset{*}{C}HO,\ Y = H_{2n}$	$1,\overset{*}{C}(X)$	< |0.2|	290
	$3,\overset{*}{C}(X)$	|1.58|	
$X = H_{2x},\ Y = \overset{*}{C}O_2H$	$1,\overset{*}{C}(Y)$	|0.39|	290
	$3,\overset{*}{C}(Y)$	|0.24|	
$X = H_{2x},\ Y = \overset{*}{C}H_2OH$	$1,\overset{*}{C}(Y)$	|0.55|	290
	$3,\overset{*}{C}(Y)$	< |0.2|	
$X = H_{2x},\ Y = \overset{*}{C}HO$	$1,\overset{*}{C}(Y)$	|0.76|	290
	$3,\overset{*}{C}(Y)$	< |0.2|	

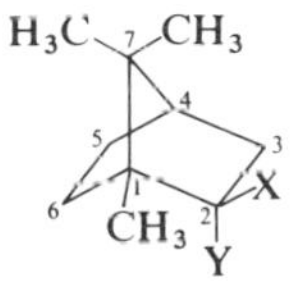

TABLE 18 (*continued*)

TABLE 18 (*continued*)

Compound[a,b]	Carbon Atom Positions	$^2J_{CC}$ (Hz)	Reference
X = $\overset{*}{C}O_2H$, Y = H$_{2n}$	1,$\overset{*}{C}$(X)	$\|0.92\|$	290
	3,$\overset{*}{C}$(X)	$\|1.44\|$	
X = $\overset{*}{C}H_2OH$, Y = H$_{2n}$	1,$\overset{*}{C}$(X)	$< \|0.25\|$	290
	3,$\overset{*}{C}$(X)	$\|0.70\|$	
X = H$_{2x}$, Y = $\overset{*}{C}O_2H$	1,$\overset{*}{C}$(Y)	$\|1.25\|$	290
	3,$\overset{*}{C}$(Y)	$< \|0.2\|$	
X = H$_{2x}$, Y = $\overset{*}{C}H_2OH$	1,$\overset{*}{C}$(Y)	$< \|0.10\|$	290
	3,$\overset{*}{C}$(Y)	$\|0.81\|$	

	Carbon Atom Positions	$^2J_{CC}$ (Hz)	Reference
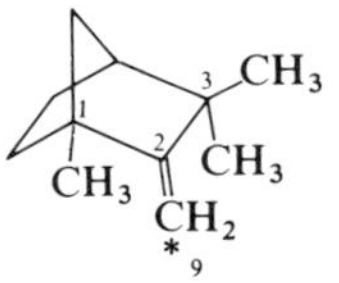	1,9	$\|1.81\|$	290
	3,9	$\|1.41\|$	

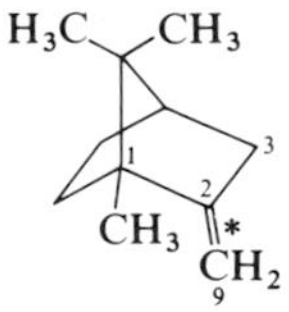	1,9	$\|1.39\|$	290
	3,9	$\|1.43\|$	

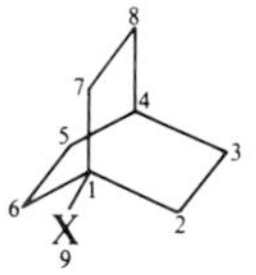			
X = $\overset{*}{C}O_2H$	2,9	$\|0.88\|$	383
X = $\overset{*}{C}H_2OH$	2,9	$\|0.58\|$	383

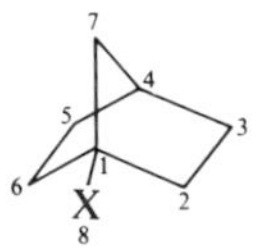			
X = $\overset{*}{C}O_2H$	2,8	$< \|0.9\|$	383
	7,8	$< \|0.9\|$	
X = CH$_2$OH	2,8	$\|1.71\|$	383
	7,8	$\|1.02\|$	

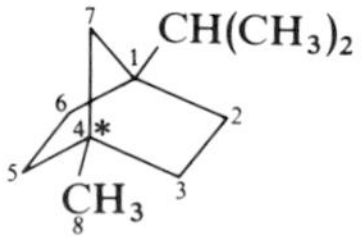	1,4	$\|6.6\|$	291

TABLE 18 (*continued*)

TABLE 18 (*continued*)

Compound[a,b]	Carbon Atom Positions	$^2J_{CC}$ (Hz)	Reference
(structure: CH(CH₃)₂, O, CH₃)	1,4 2,4	|2.4| |1.0|	291
(structure: CO₂Et, CH₃)	2,4	|1.7|	291
(structure: CO₂Et, CH₃)	2,4 4,6	|1.3| |2.2|	291

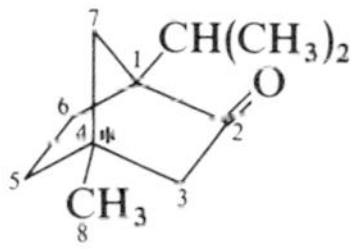

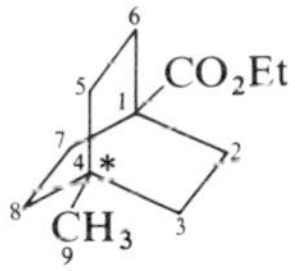

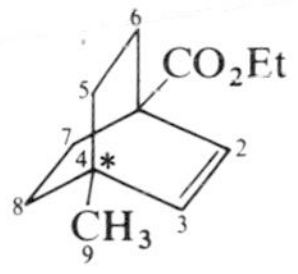

[a] See Table 10 for structures **96–108**.
[b] Asterisk indicates position of specific ^{13}C labeling.

TABLE 19. $^2J_{FC}$ VALUES IN RIGID BICYCLIC RING SYSTEMS RELATED TO NORBORNANE AND BICYCLO[2.2.2]OCTANE

Compound	Fluorine Atom Position	Carbon Atom Position	$^2J_{FC}$ (Hz)	Reference
(structure: F₂ₓ, H₂ₙ)	2x 2x	1 3	|20.2| |20.4|	80[a]
(structure: anti/syn, R, exo/endo, F₂ₓ, F₂ₙ)				
R = H	2x,2n 2x,2n	1 3	|24.1, 22.0| |24.3, 22.0|	80[a]
R = 1-CH₃	2x,2n 2x,2n	1 3	|21.9, 21.9| |24.7, 22.7|	
R = exo-3-CH₃	2x,2n 2x,2n	1 3	|24.0, 21.8| |24.0, 20.3|	
R = endo-3-CH₃	2x,2n 2x,2n	1 3	|24.2, 22.4| |22.4, 22.1|	
R = exo-5-CH₃	2x,2n 2x,2n	1 3	|23.0, 21.8| |23.6, 21.6|	

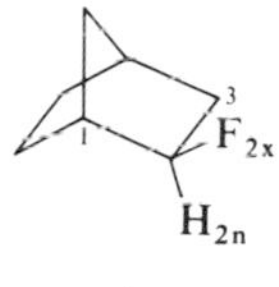

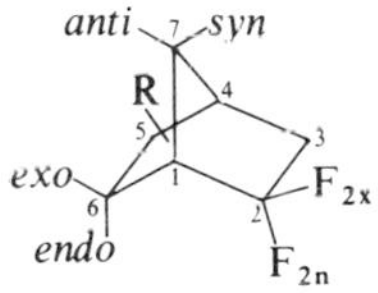

TABLE 19 (*continued*)

TABLE 19 (*continued*)

Compound	Fluorine Atom Position	Carbon Atom Position	$^2J_{FC}$ (Hz)	Reference
R = *endo*-5-CH$_3$	2x,2n	1	|23.5, 21.3|	
	2x,2n	3	|24.6, 21.8|	
R = *exo*-6-CH$_3$	2x,2n	1	|22.3, 20.9|	
	2x,2n	3	|23.1, 21.1|	
R = *endo*-6-CH$_3$	2x,2n	1	|22.3, 19.1|	
	2x,2n	3	|21.8, 21.8|	
R = *syn*-7-CH$_3$	2x,2n	1	|20.7, 20.7|	
	2x,2n	3	|25.4, 21.6|	
R = *anti*-7-CH$_3$	2x,2n	1	|22.6, 20.6|	
	2x,2n	3	|24.1, 21.7|	
	1	2	|20.3|	277
	1	7	|18.5|	

[a] $^2J_{CF}$ values are considered to be accurate to ± 0.5 Hz.[80]

TABLE 20. $^2J_{NC}$ VALUES IN RIGID BICYCLIC RING SYSTEMS RELATED TO NORBORNANE AND BICYCLO[2.2.2]OCTANE

Compound[a]	Nitrogen Atom Position	Carbon Atom Position	$^2J_{NC}$ (Hz)	Reference
109	2n	1	ca. 0[b]	293
	2n	3	ca. 0[b]	
110	2n	3	< |0.3|[b]	293
111	2n	3	ca. 0[b]	293
112	1	3	< |0.2|	294
113	1	3	< |0.2|	294
114	O=C—^{15}NH$_2$	2	|8.2|[c]	295
115	O=C—^{15}NH$_2$	2	|7.8|[c]	295
116	O=C—^{15}NH$_2$	2	|7.4|[c]	295
117	O=C—^{15}NH$_2$	2	|6.9|[c]	295
	2	4	|5.1|	296
	2	4	|2.5|	296

[a] See Table 11 for structures **109–117**.
[b] These $^2J_{NC}$ values were estimated from line width measurements.[293]
[c] These $^2J_{NC}$ values are predicted to have a negative absolute sign.[295]

TABLE 21. $^2J_{PC}$ Values in Rigid Bicyclic Ring Systems Related to Norbornane and Bicyclo[2.2.2]octane

Compound[a]	Phosphorus Atom Position	Carbon Atom Position	$^2J_{PC}$ (Hz)	Reference
X = Cl$_2$P	7	1,4	|14.6|	111
X = Me$_2$P	7	1,4	|11.0|	
X = Me$_2$(S)P	7	1,4	0.	
X = Me$_3$P$^+$ I$^-$	7	1,4	0.	
X = Cl$_2$P	7	1,4	|13.4|	111
X = Me$_2$P	7	1,4	|12.|	
	1	4	|35.|	306
	1	3,5	|4.|	
	1	3,5,8	|5.|	308, 309
	2n	1	|1.0|	310
	2n	3	|3.1|	
	10	2	|4.2|	384
	10	9	|28.1|	
	6	9	|3.0|	
	6	4	|24.5|	

TABLE 21 (*continued*)

TABLE 21 (*continued*)

Compound[a]	Phosphorus Atom Position	Carbon Atom Position	$^2J_{PC}$ (Hz)	Reference
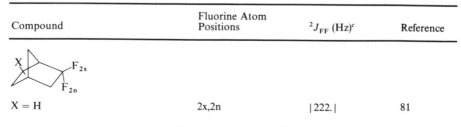	10	2	|20.1|	384
	10	9	|3.8|	
	6	9	|4.2|	
	6	4	|22.5|	
	7a	1	|3.5|	307
	8s	4	|5.|	
	2x	1	|3.8|	311
	2x	3	|1.0|	
	3s	2,4	|1.2|	311
	3a	2,4	|1.1|	311

[a] The absolute sign of $^2J_{PC}$ is negative in $(H_3CO)P(O)CH_3$.[312] The sign of $^2J_{PC}$ in $C-C-P$ systems (trivalent phosphorus) can be reversed by the presence of electronegative substituents.[385]

TABLE 22. $^2J_{FF}$ VALUES IN RIGID BICYCLIC RING SYSTEMS RELATED TO NORBORNANE AND BICYCLO[2.2.2]OCTANE[a,b]

Compound	Fluorine Atom Positions	$^2J_{FF}$ (Hz)[c]	Reference
X = H	2x,2n	|222.|	81

TABLE 22 (*continued*)

TABLE 22 (*continued*)

Compound	Fluorine Atom Positions	$^2J_{FF}$ (Hz)^c	Reference
X = 1-CH$_3$	2x,2n	|222.|	
X = *exo*-3-CH$_3$	2x,2n	|223.|	
X = *endo*-3-CH$_3$	2x,2n	|226.|	
X = 3,3-dimethyl	2x,2n	|224.|	
X = *endo*-5-CH$_3$	2x,2n	|221.|	
X = *endo*-6-CH$_3$	2x,2n	|228.|	
X = 7,7-dimethyl	2x,2n	|228.|	
X = *anti*-7-methyl	2x,2n	|222.|	

Compound	Fluorine Atom Positions	$^2J_{FF}$ (Hz)^c	Reference
R = W = CH$_3$, S = X = Y = Z = F	5x,5n	|226.0|	378
	7a,7s	|196.0|	
R = I, S = CH$_3$, W = H$_{2x}$,	5x,5n	|220.|	378
X = OCH$_3$, Y = Z = F	6x,6n	|246.|	
	7a,7s	|243.|	
R = CH$_3$, S = Y = Z = F,	5x,5n	|241.|	378
W = OCH$_3$, X = H$_{3x}$	6x,6n	|248.|	
R = I, S = Y = Z = F	5x,5n	|238.|	378
W = X = Cl	6x,6n	|223.|	
	7a,7s	|231.|	
R = W = X = Cl, S = Y = Z = F	5x,5n	|239.|	378
	6x,6n	|237.|	
	7a,7s	|232.|	

Compound	Fluorine Atom Positions	$^2J_{FF}$ (Hz)^c	Reference
X = H$_4$	5x,5n	|230.0|	378
	7a,7s	|203.0|	
X = Br	5x,5n	|220.0|	
	6x,6n	|223.5|	
	7a,7s	|197.0|	
X = I	5x,5n	|218.0|	
	6x,6n	|220.0|	
	7a,7s	|192.0|	

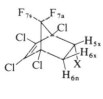

TABLE 22 (*continued*)

TABLE 22 (*continued*)

Compound	Fluorine Atom Positions	$^2J_{FF}$ (Hz)^c	Reference
X = CN	7a,7s	+174.	339
X = CO$_2$H	7a,7s	+174.	
X = Ph	7a,7s	+169.	
X = Cl	7a,7s	+178.	
X = OH	7a,7s	+182.	
X = OAc	7a,7s	+179	

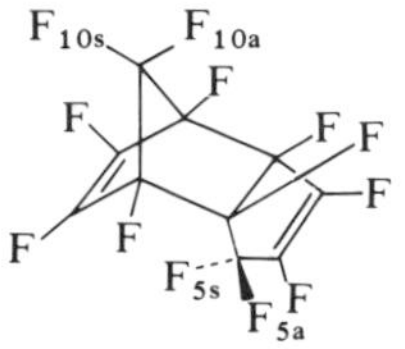	10a,10s	|185.|	386
	5a,5s	|266.|	

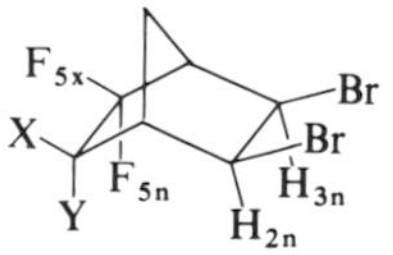 X = Y = F	5x,5n	|241.|	363
X = Y = CF$_3$	5x,5n	|255.|	
X = Y = Cl	5x,5n	|224.|	
X = H$_{6x}$, Y = F$_{6n}$	5x,5n	|248.|	
X = F$_{6x}$, Y = H$_{6n}$	5x,5n	|251.|	

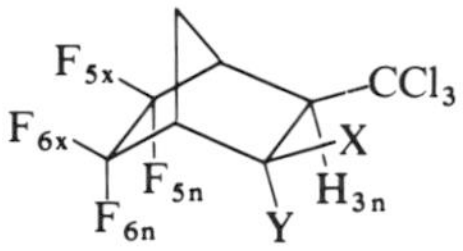 X = Cl, Y = H$_{2n}$	5x,5n	|228.|	364
X = Br, Y = H$_{2n}$	5x,5n	|228.|	
X = H$_{2x}$, Y = Cl	5x,5n	|241.|	
	6x,6n	|226.|	
X = H$_{2x}$, Y = Br	5x,5n	|240.|	
	6x,6n	|230.|	

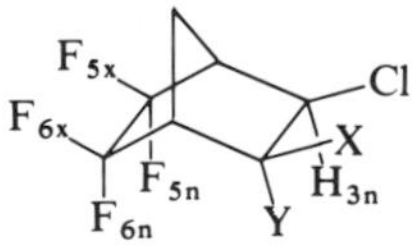 X = Cl, Y = H$_{2n}$	5x,5n	|239.|	365
	6x,6n	|239.|	
X = H$_{2x}$, Y = Cl	5x,5n	|241.|	365
	6x,6n	|244.|	
X = Y = Cl	6x,6n	|244.|	365

TABLE 22 (*continued*)

TABLE 22 (*continued*)

Compound	Fluorine Atom Positions	$^2J_{FF}$ (Hz)^c	Reference

$X = H_{6x}$, $Y = F_{6n}$	5x,5n	$\lvert 234. \rvert$	363
$X = F_{6x}$, $Y = H_{6n}$	5x,5n	$\lvert 234. \rvert$	363

^{a} An extensive tabulation of $^2J_{FF}$ values in rigid bicyclic systems is presented in ref. 275.
^{b} "The dependence of $^2J_{FF}$ on the F–C–F bond angle is quite different from that of $^2J_{HH}$"; see ref. 387.
^{c} The absolute sign of geminal $^2J_{FF}$ coupling constants in saturated systems has been shown to be positive.[388]

TABLE 23. MISCELLANEOUS $^2J_{XY}$ VALUES IN RIGID BICYCLIC SYSTEMS RELATED TO NORBORNANE AND BICYCLO[2.2.2]OCTANE

Compound	X Atom Position	Y Atom Position	$^2J_{XY}$ (Hz)	Reference
	$^{199}Hg_{2x}$ $^{199}Hg_{2x}$	$^{13}C_1$ $^{13}C_3$	$\lvert 41. \rvert$ $\lvert 43. \rvert$	317
	$^{199}Hg_{2n}$ $^{199}Hg_{2n}$	$^{13}C_1$ $^{13}C_3$	$\lvert 41. \rvert$ $\lvert 53. \rvert$	317
	$^{199}Hg_{2x}$ $^{199}Hg_{2x}$	$^{13}C_1$ $^{13}C_3$	$\lvert 57. \rvert$ $\lvert 147. \rvert$	317
	$^{199}Hg_{3x}$ $^{199}Hg_{3x}$	$^{13}C_2$ $^{13}C_4$	$\lvert 54. \rvert$ $\lvert 52. \rvert$	317

$R = H_{3x}$, $R' = H_{3n}$	$^{203/205}Tl_{5x}$	$^1H_{5n}$	$\lvert 1200. \pm 5. \rvert$	389
$R = H_{3x}$, $R' = CO_2H$	$^{203/205}Tl_{5x}$	$^1H_{5n}$	$\lvert 1197. \pm 5. \rvert$	
$R = CO_2H$, $R' = H_{3n}$	$^{203/205}Tl_{5x}$	$^1H_{5n}$	$\lvert 1204. \pm 5. \rvert$	

TABLE 23 (*continued*)

TABLE 23 (*continued*)

Compound	X Atom Position	Y Atom Position	$^2J_{XY}$ (Hz)	Reference

Compound	X Atom Position	Y Atom Position	$^2J_{XY}$ (Hz)	Reference
$L = H_{3x}$, $M = H_{3n}$, $N = H_{2x}$	$^{205}Tl_{5x}$	$^{13}C_4$	|218.|	318
	$^{205}Tl_{5x}$	$^{13}C_6$	|64.|	
$L = H_{3x}$, $M = CO_2Me$, $N = H_{2x}$	$^{205}Tl_{5x}$	$^{13}C_4$	|153.|	318
	$^{205}Tl_{5x}$	$^{13}C_6$	|90.|	
$L = H_{3x}$, $M = CO_2Me$, $N = CH_3$	$^{205}Tl_{5x}$	$^{13}C_4$	|161.|	318
	$^{205}Tl_{5x}$	$^{13}C_6$	|64.|	

| | $^{205}Tl_{2x}$ | $^1H_{2n}$ | \|797.\| | 50 |

Compound	X Atom Position	Y Atom Position	$^2J_{XY}$ (Hz)	Reference
$R = R' = H$, $Z = Tl(OAc)_2$	$^{205}Tl_{5x}$	$^{13}C_4$	|244.|[a]	318
	$^{205}Tl_{5x}$	$^{13}C_4$	|215.|[b]	319
	$^{205}Tl_{5x}$	$^{13}C_6$	|569.|[a]	318
	$^{205}Tl_{5x}$	$^{13}C_6$	|554.|[b]	319
$R = R' = H$, $Z = HgOAc$	$^{199}Hg_{5x}$	$^{13}C_4$	|43.|	319
	$^{199}Hg_{5x}$	$^{13}C_6$	|119.|	319
$R = H$, $R' = OMe$, $Z = Tl(OAc)_2$	$^{205}Tl_{5x}$	$^{13}C_4$	|232.|	318
	$^{205}Tl_{5x}$	$^{13}C_6$	|569.|	
$R = H$, $R' = Cl$, $Z = Tl(OAc)_2$	$^{205}Tl_{5x}$	$^{13}C_4$	|242.|	318
	$^{205}Tl_{5x}$	$^{13}C_6$	|566.|	
$R = Cl$, $R' = H$, $Z = Tl(OAc)_2$	$^{205}Tl_{5x}$	$^{13}C_4$	|230.|	318
	$^{205}Tl_{5x}$	$^{13}C_6$	|566.|	

Compound	X Atom Position	Y Atom Position	$^2J_{XY}$ (Hz)	Reference
$R = HgOAc$	$^{199}Hg_{5x}$	$^{13}C_4$	|53.|	321
	$^{199}Hg_{5x}$	$^{13}C_6$	|141.|	
$R = Tl(OAc)_2$	$^{205}Tl_{5x}$	$^{13}C_4$	|311.|	321
	$^{205}Tl_{5x}$	$^{13}C_6$	|681.|	

TABLE 23 (*continued*)

TABLE 23 (continued)

Compound	X Atom Position	Y Atom Position	$^2J_{XY}$ (Hz)	Reference
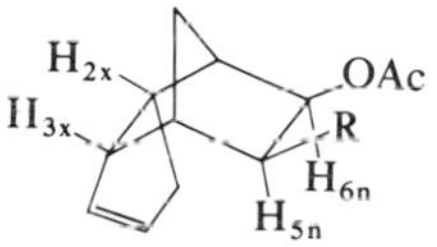				
R = Tl(OAc)$_2$	^{205}Tl$_{5x}$	^{13}C$_6$	|681.|	321
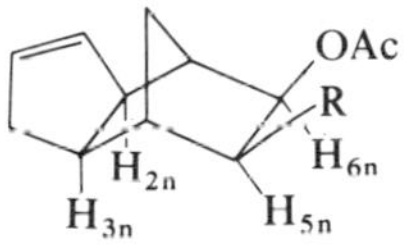				
R = HgOAc	^{199}Hg$_{5x}$	^{13}C$_4$	|48.|	321
	^{199}Hg$_{5x}$	^{13}C$_6$	|141.|	
R = Tl(OAc)$_2$	^{205}Tl$_{5x}$	^{13}C$_4$	|304.|	321
	^{205}Tl$_{5x}$	^{13}C$_6$	|676.|	
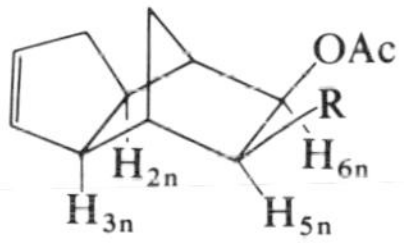				
R = HgOAc	^{199}Hg$_{5x}$	^{13}C$_6$	|143.|	321
R = Tl(OAc)$_2$	^{205}Tl$_{5x}$	^{13}C$_6$	|674.|	321
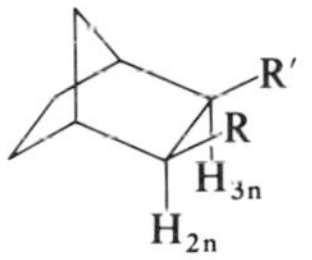				
R = Tl(OAc)$_2$, R′ = OAc	$^{203/205}$Tl$_{2x}$	^{13}C$_1$	|322.3|	320
	$^{203/205}$Tl$_{2x}$	^{13}C$_3$	|683.6|	
R = HgOAc, R′ = OAc	^{199}Hg$_{2x}$	^{13}C$_1$	|51.8|	320
	^{199}Hg$_{2x}$	^{13}C$_3$	|142.9|	
R = HgOAc, R′ = H$_{3x}$	^{199}Hg$_{2x}$	^{13}C$_1$	|41.|	320
	^{199}Hg$_{2x}$	^{13}C$_3$	|43.|	
R = HgCl, R′ = OAc	^{199}Hg$_{2x}$	^{13}C$_1$	|57.0|	320
	^{199}Hg$_{2x}$	^{13}C$_3$	|147.0|	
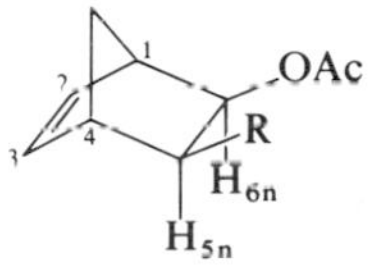				
R = Tl(OAc)$_2$	$^{203/205}$Tl$_{5x}$	^{13}C$_4$	|239.3|	320
	$^{203/205}$Tl$_{5x}$	^{13}C$_6$	|427.3|	

TABLE 23 (continued)

TABLE 23 (*continued*)

Compound	X Atom Position	Y Atom Position	$^2J_{XY}$ (Hz)	Reference
R = HgOAc	$^{199}Hg_{5x}$	$^{13}C_4$	$\lvert 40.3 \rvert$	320
	$^{199}Hg_{5x}$	$^{13}C_6$	$\lvert 96.4 \rvert$	

R = Tl(OAc)$_2$	$^{203/205}Tl_{5x}$	$^{13}C_4$	$\lvert 219.7 \rvert$	320
	$^{203/205}Tl_{5x}$	$^{13}C_6$	$\lvert 61.0 \rvert$	
R = HgOAc	$^{199}Hg_{5x}$	$^{13}C_4$	$\lvert 39.1 \rvert$	320
R = HgCl	$^{199}Hg_{5x}$	$^{13}C_4$	$\lvert 36.6 \rvert$	320

	$^{119}Sn_{2x}$	$^{13}C_1$	$\lvert 40.3 \rvert$	316
	$^{119}Sn_{2x}$	$^{13}C_3$	$\lvert 37.6 \rvert$	

	$^{119}Sn_{2n}$	$^{13}C_1$	$\lvert 40.9 \rvert$	316
	$^{119}Sn_{2n}$	$^{13}C_3$	$\lvert 33.8 \rvert$	

	$^{77}Se_2$	$^{13}C_3$	$\lvert 17.8 \rvert$	268

[a] CDCl$_3$ solvent.[318]
[b] Pyridine solvent.[319]

Three-Bond (Vicinal) Coupling (^3J)

Vicinal proton–proton couplings ($^3J_{HH}$) have been extensively utilized in conformational analysis. Their usefulness in this regard stems principally from the "Karplus relation,"[390] which relates the magnitude of $^3J_{AB}$ in the (saturated) system H_A—C—C—H_B to the dihedral angle ϕ between H_A and H_B in **121**. For protons H_A and H_B in this system, $^3J_{AB} = A \cos^2 \phi + B \cos \phi + C$, where ϕ is defined in the Newman projection **121**. Since its first introduction, application of the Karplus relation has been extended to permit correlation of a number of vicinal coupling constants $^3J_{XY}$ with dihedral angle ϕ across a variety of systems X—L—M—Y where L and M represent sp^2- or sp^3-hybridized carbon atoms or heteroatoms. A listing of some representative

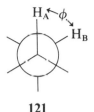

121

systems in which the angular dependence of vicinal couplings constants has been studied is presented in Scheme 10. Uses (and abuses) of the Karplus relation in conformational and configurational analysis have been examined in a number of penetrating reviews.[286,324,325,419,420]

Because of their well-defined molecular geometry, rigid bicyclic systems (e.g., norbornanes and bicyclo[2.2.2]octanes) have been extensively utilized as substrates for demonstrating the existence (or nonexistence) of dihedral angular dependencies of vicinal couplings, $^3J_{XY}$. In this section, we will consider briefly some of the conformational (geometric) and substituent (electronic and steric) effects on $^3J_{XY}$ values that have been studied specifically in norbornyl and bicyclo[2.2.2]octyl systems.

A number of vicinal proton–proton coupling constants that have been measured in rigid bicyclic ring systems related to norbornane and bicyclo-[2.2.2]octane are presented in Table 24. In these systems, the dihedral angle dependence appears to be the major factor affecting the magnitude of $^3J_{HH}$. However, significant deviations from the Karplus relation can occur in unsymmetrically substituted norbornanes and norbornenes. In such cases, torsional effects can result that distort these ring systems, leading to larger variations in $^3J_{HH}$ than are expected on the basis of a straightforward dihedral angle dependence in the unsubstituted (and undistorted) ring system.[356,433,434]

Even in unsubstituted or symmetrically substituted norbornanes, it is evident that factors other than dihedral angle indeed contribute to $^3J_{HH}$. For example, the exo,exo and endo,endo vicinal proton–proton couplings in norbornenes (**122**) are virtually identical,[63,333,340] whereas in symmetrically substituted[341] and in unsubstituted[335] norbornanes (**123**) the exo,exo vicinal proton–proton coupling (ca. 12 Hz) is generally larger than the corresponding endo,endo coupling (ca. 9 Hz). When viewed solely in terms of the Karplus relation, this nonequivalence observed in norbornanes is surprising, because exo,exo and endo,endo dihedral angles are about the same (ca. 0°) in **123** (as they are also in **122**). Recent theoretical studies on derivatives of these systems

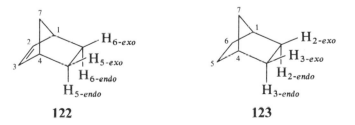

        122   **123**

Scheme 10. *Systems* X–L–M–Y *in Which Angular Dependence of* ^3J *Spin Couplings Has Been Demonstrated*

System	Observations	Leading references
$^1H-C-C-^1H$	"Karplus relation" $[^3J = f(\phi)]$ was first established for $^3J_{HCCH}$.[323,390,391]	323, 391, 392
$^1H-C-N-^1H$	$^3J_{HCNH} = 9.4 \cos^2 \phi - 1.1 \cos \phi + 0.4$ (in peptides).	393
$^1H-C-\overset{+}{N}-^1H$	"The maximum value of gauche $^3J_{HCN^+H}$ is 3.15 Hz and the minimum value of trans $^3J_{HCN^+H}$ is 9.45 Hz" in a series of substituted protonated benzylamines.	394
$^1H-C-O-^1H$	Angular dependence of $^3J_{HCOH}$ parallels the Karplus relation[279a] for $^3J_{HCCH}$.[316] $^3J_{HCOH} = 10.6 \cos^2 \phi - 1.4 \cos \phi$.[316]	395–397
$^{19}F-C-C-^1H$	$^3J_{FCCH} = \begin{cases} 31 \cos^2 \phi (0 \leq \phi \leq 90°)^{376} \\ 44 \cos^2 \phi (90° \leq \phi \leq 180°)^{376} \end{cases}$ A Karplus relation has been calculated for $^3J_{FCCH}$ in FCH_2-CH_2F and FCH_2-CF_3 for a variety of FCCH dihedral angles.[387]	376, 377, 387, 398, 399
$^{31}P-C-C-^1H$	$^3J_{PCCH} = \begin{cases} 18 \cos^2 \phi (0 \leq \phi \leq 90°) \\ 41 \cos^2 \phi (90° \leq \phi \leq 180°) \end{cases}$ for $RP(O)(OCH_3)_2$ compounds.[338]	338, 380, 382, 400
$^{31}P-O-C-^1H$	Separate $^3J_{POCH}$ correlations with dihedral angle are required for trivalent and pentavalent phosphorous systems. The "partial angular dependence" of $^3J_{POCH}$ couplings in 2-oxo-1,3,2-dioxaphosphor-inane derivatives has been reported.	401–403
$^{15}N-C-C-^1H$	The results of INDO molecular orbital calculations suggest that "$^3J_{NCCH}$ depends both on the orientation of the nitrogen lone pair and on the HCCN dihedral angle."[404]	382, 404
$^{14}N^+-C-C-^1H$	$^3J_{NCCH}^+ = (1.41 \pm 0.08) - (0.56 \pm 0.13) \cos \phi + (1.51 \pm 0.12) \cos 2\phi$.[299]	299, 405
$^{13}C-C-C-^1H$	$^3J_{CCCH} = 4.26 - 1.00 \cos \phi + 3.56 \cos 2\phi$.[407]	375, 382, 406, 407
$^{13}C-O-C-^1H$	"Vicinal $^{13}C-^1H$ coupling, both through C—C and C—O bonds, shows an orientational dependence analogous to that for protons."[406]	406
$^{13}C-N-C-^1H$	A Karplus-type relationship was found for $^3J_{CNCH}$ values measured in uridine derivatives.[408]	408
$^{13}C-C-C-^{13}C$	The Karplus relation "does not provide a completely adequate quantitative representation" of $^3J_{CCCC}$.[285] Non-bonded interactions are substantial for coupling to ^{13}C and can lead to	286, 285, 291, 409–411

Scheme 10 (*continued*)

Scheme 10 (*continued*)

System	Observations	Leading references
	important deviations from angular dependencies of the Karplus type."[409,411] However, a Karplus-type relationship was found for $^3J_{CCCC}$ values measured in similarly substituted rigid bicyclic systems bearing site-specific skeletal ^{13}C atom enrichment: $^3J_{CCCC} = 1.67 + 0.176 \cos \phi + 2.24 \cos 2\phi$.[291]	
$^{13}C-C-O-^{31}P$	$^3J_{CCOP}$ couplings in polyuridylic acid, uridine, and related nucleotides in aqueous solution appear to be "sensitive to the relative populations of the rotamers about the corresponding C—O bonds."[413]	412, 413
$^{12}C-C-C-^{31}P$	A "highly asymmetric dihedral dependence of $^3J_{CCCP}$" was noted in $RP(O)(OCH_3)_2$ compounds.[313] Thiem and Meyer report $^2J_{CCCP} = (7.35 \pm 0.25) - (1.76 \pm 0.39) \cos \phi + (7.86 \pm 0.32) \cos 2\phi$ in rigid camphene phosphonates.[315] "Separate correlations" are required for P(III) and P(V) compounds.[401] "The origin of the unique shape of the P(III) $^3J_{CCCP}$ vs. ϕ plot is tentatively associated with the presence of the lone pair on phosphorus, through the adoption of preferred conformations."[414] Karplus plots with minima located near $\phi = 90°$ have been obtained for P(IV) compounds (phosphonates, phosphine oxides, phosphine sulfides, and phosphonium halides); however, the dihedral angle dependence of $^3J_{CCCP}$ in P(III) compounds (phosphonites, phosphonous dihalides, and phosphines) is not adequately described by the simple Karplus relationship.[414]	308, 311, 315, 401, 414
$^{13}C-C-C-^{14}N^+$	A Karplus-type angular dependence for $^3J_{CCCN^+}$ was observed.[293]	293, 294, 415, 416
$^{13}C-C-C-^{205}Tl$	"A Karplus-type of dihedral angle dependence of $^3J_{CCCTl}$" was noted.[319]	319
$^{13}C-C-C-^{119}Sn$	$^3J_{CCCSn} = 30.4 - 7.6 \cos \phi + 25.2 \cos 2\phi$; mean deviation 3.75 Hz.[316]	316
$^{13}C-C-C-^{199}Hg$	A plot of $^3J_{CCCHg}$ vs. dihedral angle afforded "clearly a curve of the Karplus form."[317]	317
$^{19}F-C-C-^{19}F$	$^3J_{FCCF}$ couplings "exhibit a complex angular dependence which includes at least one sign inversion."[418]	387, 401, 417, 418

TABLE 24. $^3J_{HH}$ Values in Rigid Bicyclic Ring Systems Related to Norbornane and Bicyclo[2.2.2]octane

Compound	Proton Positions	$^3J_{HH}$ (Hz)[a]	References

Compound	Proton Positions	$^3J_{HH}$ (Hz)[a]	References
$W = H_{5x}$, $X = H_{5n}$,	1,7s	|1.8|	63, 264
$Y = H_{6x}$, $Z = H_{6n}$	1,7s	|2.0–2.2|	125
	1,7a	|1.5|	63,264
	1,7a	|1.5–1.6|	125
	1,6x	|3.5|	63
	1,6n	0.0	63,125
	5x,6x	$+9.30 \pm 0.05$	335
	5n,6n	$+9.02 \pm 0.05$	335
	5x,6n	|4.4|	63
	5x,6n	$+3.87 \pm 0.05$	335
$W = H_{5x}$, $X = H_{6x}$,	1,7s	|2.0|	258
$Y = Z = Cl$	1,7a	$< \|1.3–1.5\|$	
	1,6x	|3.2|	
	5x,6x	|7.5|	
	1,2 = 3,4	|2.85|	
	2,3	|5.55|	
$W = Z = Cl$, $X = H_{6x}$,	1,7s	|2.0|	258
$Y = H_{5n}$	1,7a	$< \|1.3–1.5\|$	
	1,6x	|3.9|	
	1,6n	0.0	
	5n,6x	|2.1|	
	1,2 = 3,4	|2.95|	
	2,3	|5.80|	
$W = H_{5x}$, $X = Z = Cl$,	1,7s	|1.8|	258
$Y = H_{5n}$	1,7a	$< \|1.3–1.5\|$	
	1,6x	|3.8|	
	1,6n	0.0	
	1,2 3,4	|3.0|	
	2,3	|5.60|	
$W = H_{5x}$, $X = H_{6x}$,	1,7s	|1.95|	258
$Y = Cl$, $Z = H_{6n}$	1,7a	$< \|1.3 − 1.5\|$	
	1,6x	|3.75|	
	4,5x	|3.6|	
	1,6n	0.0	
	5x,6x	|8.2|	
	5x,6n	|3.4|	
	1,2	|2.80|	
	3,4	|3.0|	
	2,3	|5.60|	

TABLE 24 (*continued*)

TABLE 24 (*continued*)

Compound	Proton Positions	$^3J_{HH}$ (Hz)^a	References
W = PhCH$_2$S=O, X = H$_{6x}$,	2,3	|6.|	422
Y = H$_{5n}$, Z = H$_{6n}$	3,4	|3.|	
W = PhS=O, X = H$_{6x}$,	2,3	|6.|	422
Y = H$_{5n}$, Z = H$_{6n}$	3,4	|3.|	
	5n,6n	|8.|	
	5n,6x	|4.|	
	1,6x	|3.|	
W = PhCH$_2$S, X = H$_{6x}$,	1,2	|3.|	422
Y = H$_{5n}$, Z = H$_{6n}$	2,3	|5.|	
	3,4	|3.|	
	5n,6n	|8.|	
	5n,6x	|4.|	
W = PhS, X = H$_{6x}$,	1,2	|3.|	422
Y = H$_{5n}$, Z = H$_{6n}$	2,3	|5.5|	
	3,4	|3.|	
	5n,6n	|8.|	
	5n,6x	|4.|	
W = H$_{5x}$, X = H$_{6x}$,	1,6x	|3.3|	329
Y = Z = Br	1,7s	|2.1|	
	1,7a	ca. |1.5|	
W = H$_{5x}$, X = H$_{6x}$,	1,7s	|1.9|	329
Y = Br, Z = Cl	5x,6x	|7.4|	
W = X = Br, Y = H$_{5n}$,	1,7s	|1.9|	329
Z = H$_{6n}$	1,7a	ca. |1.7|	
W = Br, X = Cl,	1,7s	|1.9|	329
Y = H$_{5n}$, Z = H$_{6n}$	5n,6n	|6.8|	
W = Z = Br, X = H$_{6x}$,	1,6x	|2.9|	329
Y = H$_{5n}$	1,7s	|2.1|	
	2,3	ca. |5.9|	
	5n,6x	|2.5|	
W = H$_{5x}$, X = Cl,	4,5x	|3.5|	329
Y = Br, Z = H$_{6n}$	5x,6n	|2.4|	
W = CO$_2$Me, X = CO$_2$H,	1,2	|1.9|	330
Y = H$_{5n}$, Z = H$_{6n}$	3,4	|1.9|	
	1,6n	ca. |0.2|	
	1,5n	ca. |0.2|	
	1,7s	|1.8|	
	4,7s	|1.8|	
	1,7a	|1.6|	
	4,7a	|1.6|	
W = H$_{5x}$, X = H$_{6x}$,	1,2	|2.7|	330
Y = CO$_2$Me, Z = CO$_2$H	3,4	|2.7|	
	1,6x	ca. |2.|	
	4,5x	ca. |2.|	
	1,7s	|1.8|	
	4,7s	|1.8|	
	1,7a	|1.4|	

TABLE 24 (*continued*)

TABLE 24 (*continued*)

Compound	Proton Positions	$^3J_{HH}$ (Hz)^a	References
	4,7a	|1.4|	
	2,3	|5.7|	
$W = X = CO_2Me,$	1,2 = 3,4	|1.9|	330
$Y = H_{5n}, Z = H_{6n}$	1,6n = 4,5n	ca. 0.	
	1,7s = 4,7s	|1.7|	
	1,7a = 4,7a	|1.6|	
$W = H_{5x}, X = H_{6x},$	1,2 = 3,4	|1.7|	330
$Y = Z = CO_2Me$	1,6x = 4,5x	|1.6|	
	1,7s = 4,7s	|1.8|	
	1,7a = 4,7a	|1.4|	
$W = Z = CO_2Me,$	1,2	|2.7|	330
$X = H_{6x}, Y = H_{5n}$	2,3	|5.6|	
	3,4	|3.1|	
	4,5n	ca. |0.6|	
	1,6x	|3.8|	
	5n,6x	|4.4|	
	1,7s = 4,7s	|1.5|	
	1,7a = 4,7a	|1.1|	
W,X 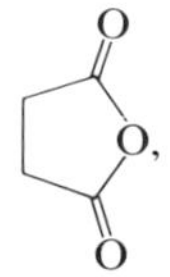 =	1,2 = 3,4	|1.9|	330
$Y = H_{5n}, Z = H_{6n}$	1,6n = 4,5n	|0.5|	
	1,7s = 4,7s	|1.7|	
	1,7a = 4,7a	|1.5|	
$W = H_{5x}, X = H_{6x},$	1,2 = 3,4	|1.8|	330
	1,6x = 4,5x	|2.2|	
Y,Z = 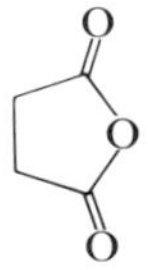	1,7s = 4,7s	|1.4|	
$W = H_{5x}, X = H_{6x},$	1,7a = 4,7a	|1.3|	423
$Y = \overset{O}{\overset{\|}{P}}(OEt)_2, Z = Cl$	1,6x	|3.4|	
	5x,6x	|7.8|	
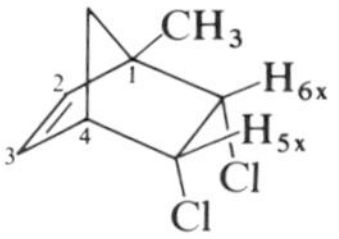	1,6x	|3.7|	329
	1,7s	|2.3|	
	2,3	|5.8|	
	3,4	|3.8|	
	5x,6x	|7.8|	

TABLE 24 (*continued*)

TABLE 24 (*continued*)

Compound	Proton Positions	$^3J_{HH}$ (Hz)^a	References
	5n,6x 3,4	|2,4| |3.6|	329
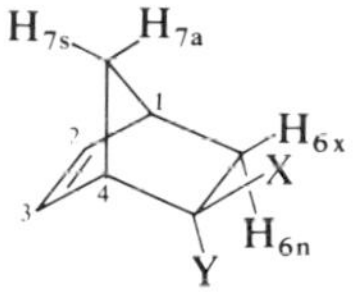			
X = H$_{5x}$, Y = OH	1,2 2,3 3,4 1,7s = 4,7s 1,7a = 4,7a 1,6x 1,6n 4,5x 5x,6x 5x,6n 1,2 3,4 2,3 1,6x	|2.9| |5.9| |2.6| |1.8| |1.4| |3.8| 0. |3.5| |8.0| |3.0| |3.| |3.| |6.| |4.|	331 332
X = OH, Y = H$_{5x}$	1,2 2,3 3,4 1,7s − 4,7s 1,7a = 4,7a 1,6x 1,6n 4,5n 5n,6n 5n,6x 1,2 3,4 2,3	|3.0| |5.8| |2.8| |1.9| |1.3| |3.7| 0. 0. |5.6| |3.1| |3.| |3.| |6.|	331 332
X = OH, Y = CH$_3$	1,2 3,4 1,6x	|2.| |2.| |4.|	332
X = CH$_3$, Y − OH	1,2 3,4 2,3	|3.| |3.| |6.|	332
X = H$_{5x}$, Y = CO$_2$Me	1,2 2,3	|2.8| |5.6|	331

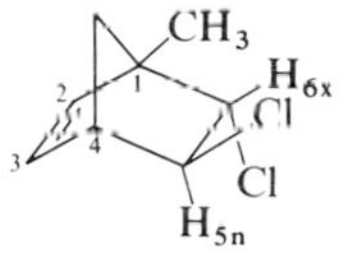

TABLE 24 (*continued*)

TABLE 24 (*continued*)

Compound	Proton Positions	$^3J_{HH}$ (Hz)^a	References
	3,4	|2.8|	
	1,7s = 4,7s	|1.9|	
	1,7a = 4,7a	|1.3|	
	1,6x	|3.4|	
	1,6n	0.	
	4,5x	|3.6|	
	5x,6x	|8.8|	
	5x,6n	|3.4|	
	5x,6x	+9.4	333
	5x,6n	+4.2	
X = CO$_2$Me, Y = H$_{5n}$	1,2	|2.7|	331
	3,4	|2.8|	
	1,7s = 4,7s	|1.8|	
	1,7a = 4,7a	|1.4|	
	1,6x	|3.5|	
	1,6n	0.	
	4,5n	0.	
	5n,6n	|4.4|	
	5n,6x	|3.8|	
X = H$_{5x}$, Y = CN	1,2	|2.6|	331
	2,3	|5.6|	
	3,4	|2.9|	
	1,7s = 4,7s	|1.9|	
	1,7a = 4,7a	|1.3|	
	1,6x	|3.6|	
	1,6n	0.	
	4,5x	|3.5|	
	5x,6x	|9.1|	
	5x,6n	|3.4|	
X = CN, Y = H$_{5n}$	1,2	|3.0|	331
	2,3	|5.8|	
	3,4	|2.7|	
	1,7s = 4,7s	|1.8|	
	1,7a = 4,7a	|1.4|	
	1,6x	|3.4|	
	1,6n	0.	
	4,5n	0.	
	5n,6n	|4.5|	
	5n,6x	|4.2|	
X = H$_{5x}$, Y = SiMe$_3$	1,2	|2.85|	334
	3,4	|2.85|	
	1,6x	|3.80|	
	1,6n	|0.5|	
	1,7s = 4,7s	|1.8|	
	1,7a = 4,7a	|1.55|	
	2,3	|5.7|	
	4,5x	|2.85|	

TABLE 24 (*continued*)

TABLE 24 (*continued*)

Compound	Proton Positions	$^3J_{HH}$ (Hz)[a]	References
	5x,6x	|9.30|	
	5x,6n	|5.55|	
X = H$_{5x}$, Y = $\overset{\overset{\displaystyle O}{\|}}{P}(OMe)_2$	5x,6x	|10.0|	379
	5x,6n	|5.2|	

X = Y = OMe, Z = $\overset{\overset{\displaystyle O}{\|}}{P}(OMe)_2$	5x,6x	|10.0|	379
	5x,6n	|5.0|	
X = H$_{7s}$, Y = Cl, Z = $\overset{\overset{\displaystyle O}{\|}}{P}(OMe)_2$	5x,6x	|10.2|	379
	5x,6n	|5.2|	
X = H$_{7s}$, Y = Cl, Z = CN	5x,6x	+9.83	354
	5x,6n	+4,27	
X = H$_{7s}$, Y = Cl, Z = CO$_2$Me	5x,6x	+9.08	354
	5x,6n	+4.29	
X = H$_{7s}$, Y = Cl, Z = OAc	5x,6x	+7.77	354
	5x,6n	+2.62	
X = H$_{7s}$, Y = Cl, Z – Br	5x,6x	+8.55	354
	5x,6n	+3.47	
X = H$_{7s}$, Y = Z = Cl	5x,6x	+8.52	354
	5x,6n	+3.20	
X = H$_{7s}$, Y = Cl, Z = Ph	5x,6x	+9.48	354
	5x,6n	+4.42	
X = Cl, Y = H$_{7a}$, Z = CN	5x,6x	+9.43	354
	5x,6n	+4.45	
X – Cl, Y = H$_{7a}$, Z – CO$_2$Me	5x,6x	+9.80	354
	5x,6n	+4.18	
X = Cl, Y = H$_{7a}$, Z = OAc	5x,6x	+8.10	354
	5x,6n	+2.83	
X = Cl, Y = H$_{7a}$, Z = Br	5x,6x	+8.96	354
	5x,6n	+4.07	
X = Z = Cl, Y = H$_{7a}$	5x,6x	+8.60	354
	5x,6n	+3.53	
X = Cl, Y = H$_{7a}$, Z = Ph	5x,6x	+8.82	354
	5x,6n	+5.61	
X = H$_{7s}$, Y = Z = OAc	5x,6x	|8.3|	355
	5x,6n	|2.8|	
X = H$_{7s}$, Y = H$_{7a}$, Z = OAc	5x,6x	|8.5|	355
	5x,6n	|3.0|	

TABLE 24 (*continued*)

TABLE 24 (*continued*)

Compound	Proton Positions	$^3J_{HH}$ (Hz)^a	References
X = Y = F, Z = OAc	5x,6x	+8.07	339
	5x,6n	+2.69	
X = Y = F, Z = OH	5x,6x	+7.94	339
	5x,6n	+2.88	
X = Y = F, Z = Cl	5x,6x	+8.94	339
	5x,6n	+3.40	
X = Y = F, Z = Ph	5x,6x	+9.28	339
	5x,6n	+4.56	
X = Y = F, Z = CO$_2$H	5x,6x	+9.70	339
	5x,6n	+4.54	
X = Y = F, Z = CN	5x,6x	+9.70	339
	5x,6n	+4.37	

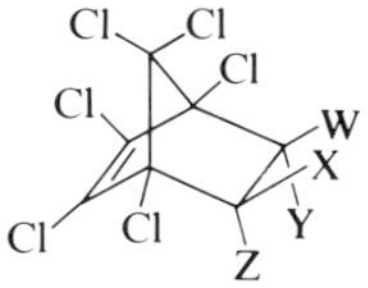 Compound	Proton Positions	$^3J_{HH}$ (Hz)^a	References
W = H$_{5x}$, X = H$_{6x}$, Y = Z = Me	5x,6x	+9.35	357
W = H$_{5x}$, X = Y = Me, Z = H$_{6n}$	5x,6n	+5.66	357
W = H$_{5x}$, X = H$_{6x}$, Y = H$_{6n}$, Z = Me	5x,6x	+8.83	357
	5n,6x	+4.02	

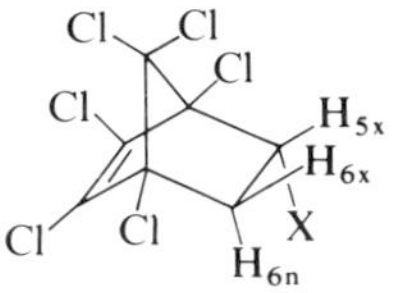

56

Compound	Proton Positions	$^3J_{HH}$ (Hz)^a	References
X = H$_{5n}$	5x,6x	+9.52 ± 0.05	335
	5n,6n	+9.10 ± 0.05	
	5x,6n	+3.70 ± 0.05	
X = CN	5x,6x	+9.3	127
	5x,6n	+4.6	
X = CO$_2$H	5x,6x	+8.5	127
	5x,6n	+4.4	
X = Ph	5x,6x	+8.9	127
	5x,6n	+4.2	
X = Cl	5x,6x	+8.0	127
	5x,6n	+3.2	
	5x,6x	+8.2	337

TABLE 24 (*continued*)

TABLE 24 (continued)

Compound	Proton Positions	$^3J_{HH}$ (Hz)[a]	References
X = OH	5x,6x	+7.4	127
	5x,6n	+2.4	
X = OAc	5x,6x	+7.6	127, 355
	5x,6n	+2.5	
$X = \overset{\overset{O}{\|\|}}{P(OMe)_2}$	5x,6x	+9.7 ± 0.10	338
	5x,6n	+5.1 ± 0.09	
	5x,6x	|9.5|	380
	5x,6n	|4.97|	

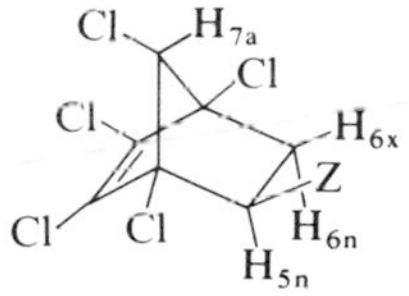			
Z = CN	5n,6n	+10.00	354
	5n,6x	+4.40	
Z = CO$_2$Me	5n,6n	+9.96	354
	5n,6x	+3.73	
Z = OAc	5n,6n	+7.84	354
	5n,6x	+2.61	
Z = Br	5n,6n	+8.30	354
	5n,6x	+3.39	
Z = Cl	5n,6n	+8.37	354
	5n,6x	+3.25	
Z = OAc	5n,6n	|7.7|	355
	5n,6x	|2.7|	

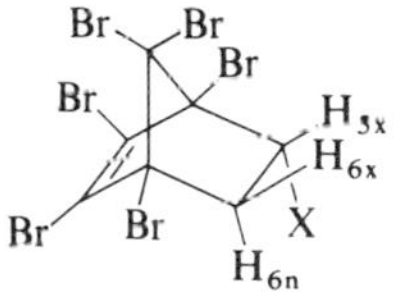

119

X = H$_{5n}$	5x,6x	+9.6	340
	5n,6n	+8.9	
	5x,6n	+3.7	
X = CN	5x,6x	+9.10	340
	5x,6n	+4.09	
X = CO$_2$Me	5x,6x	+8.78	340
	5x,6n	+4.03	
X = Ph	5x,6x	+9.06	340
	5x,6n	+4.19	
X = OAc	5x,6x	+7.56	340
	5x,6n	+2.45	

TABLE 24 (continued)

TABLE 24 (*continued*)

Compound	Proton Positions	$^3J_{HH}$ (Hz)^a	References
R = S = OMe, W = X = Y = Z = H	5x,6x	$+10.76 \pm 0.05$	335
	5n,6n	$+9.80 \pm 0.05$	
	5x,6n	$+4.56 \pm 0.05$	
R,S = O, W = X = Y = Z = H	5x,6x	$+11.35 \pm 0.05$	335
	5n,6n	$+11.35 \pm 0.05$	
	5x,6n	$+4.90 \pm 0.05$	
R = H$_{7s}$, S = W = Br,	1,6x	$\lvert 3.5 \rvert$	362
X = H$_{6x}$, Y = H$_{5n}$,	1,7s	$\lvert 1.2–1.4 \rvert$	
Z = H$_{6n}$	4,7s	$\lvert 1.2–1.4 \rvert$	
	5n,6x	$\lvert 4.5 \rvert$	
	5n,6n	$\lvert 8. \rvert$	
R = H$_{7s}$, S = H$_{7a}$,	1,7a	$\lvert 1.4 \rvert$	362
W = H$_{5x}$, X = Y = Cl,	4,7a	$\lvert 1.4 \rvert$	
Z = H$_{6n}$	1,7s	$\lvert 1.8 \rvert$	
	4,7s	$\lvert 1.8 \rvert$	
	4,5x	$\lvert 4.8 \rvert$	
	5x,6n	$\lvert 2.4 \rvert$	
R = W = Cl, S = H$_{7a}$,	1,7a	$\lvert 1.7 \rvert$	362
X = H$_{6x}$, Y = H$_{5n}$,	4,7a	$\lvert 1.7 \rvert$	
Z = H$_{6n}$	5n,6x	$\lvert 4.3 \rvert$	
	5n,6n	$\lvert 7.0 \rvert$	
R = H$_{7s}$, S = Y = Cl,	1,6x	$\lvert 4.0 \rvert$	362
W = H$_{5x}$, X = H$_{6x}$,	1,7s	$\lvert 1.7–1.8 \rvert$	
Z = H$_{6n}$	4,7s	$\lvert 1.7–1.8 \rvert$	
	4,5x	$\lvert 3.7 \rvert$	
	5x,6x	$\lvert 9. \rvert$	
	5x,6n	$\lvert 3.7 \rvert$	
R = H$_{7s}$, S = H$_{7a}$,	1,7a	$\lvert 1.6 \rvert$	362
W = X = Cl, Y = H$_{5n}$,	4,7a	$\lvert 1.6 \rvert$	
Z = H$_{6n}$	1,7s	$\lvert 1.6 \rvert$	
	4,7s	$\lvert 1.6 \rvert$	
R = H$_{7s}$, S = W = Cl,	1,6x	$\lvert 4.0–4.5 \rvert$	362
X = H$_{6x}$, Y = H$_{5n}$,	1,7s	$\lvert 1.3–1.4 \rvert$	
Z = H$_{6n}$	4,7s	$\lvert 1.3–1.4 \rvert$	
	5n,6n	$\lvert 8. \rvert$	
	5n,6x	$\lvert 4.5 \rvert$	
R = Cl, S = H$_{7a}$,	1,7a	$\lvert 1.7 \rvert$	362
W = OAc, X = H$_{6x}$,	4,7a	$\lvert 1.7 \rvert$	
Y = H$_{5n}$, Z = H$_{6n}$	5n,6n	$\lvert 6.6 \rvert$	
	5n,6x	$\lvert 4.2 \rvert$	
R = H$_{7s}$, S = Cl,	1,6x	$\lvert 3.7 \rvert$	362
W = OAc, W = H$_{6x}$,	1,7s	$\lvert 1.3–1.4 \rvert$	
Y = H$_{5n}$, Z = H$_{6n}$	4,7s	$\lvert 1.3–1.4 \rvert$	

TABLE 24 (*continued*)

TABLE 24 (*continued*)

Compound	Proton Positions	$^3J_{HH}$ (Hz)^a	References
	5n,6n	|7.6|	
	5n,6x	|3.7|	
R = Cl, S = H$_{7a}$,	1,6x	|3.6–3.7|	362
W = Br, X = H$_{6x}$,	1,7a	|1.6–1.7|	
Y = H$_{5n}$, Z = H$_{6n}$	4,7a	|1.6–1.7|	
	5n,6n	|8.0|	
	5n,6x	|3.7–3.8|	

Compound	Proton Positions	$^3J_{HH}$ (Hz)^a	References
122a X = H$_{7s}$, Y = OH,	1,6	|2.–3.5|	362
Cl, or Br, Z = H$_5$	1,7a	|1.7–2.|	
122b X = H$_{7s}$, Y = H$_{7a}$,	4,7a	|1.7–2.|	
Z = H$_5$ or Cl	1,7s	|1.7–2.|	
122c X = Cl, Y = H$_{7a}$,	4,7s	|1.7–2.|	
Z = H$_5$			

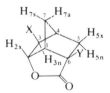

Compound	Proton Positions	$^3J_{HH}$ (Hz)^a	References
X = H$_{3x}$, Y = H$_{6x}$	1,2x	|5.2|	424
	1,6x	|3.6|	
	1,7a	|1.4|	
	1,7s	|1.4|	
	2x,3x	|7.6|	
	5x,6x	|9.1|	
	5n,6x	|3.6|	
X = I, Y = H$_{6x}$	1,2x	|5.4|	356
	1,6x	|5.0|	
	3n,4	|0.5|	
	4,5x	|3.6|	
	4,5n	|0.5|	
	5n,6x	|3.0|	
	5x,6x	|10.2|	
	1,7s	|1.8|	
	4,7s	|1.6|	
	1,7a	|0.6|	
	4,7a	|1.5|	

TABLE 24 (continued)

TABLE 24 (*continued*)

Compound	Proton Positions	$^3J_{HH}$ (Hz)^a	References
X = OAc, Y = H$_{6x}$	1,2x	|4.6|	356
	1,6x	|4.6|	
	3n,4	|0.8|	
	4,5x	|3.8|	
	5x,6x	|10.5|	
	1,7s	|1.5|	
	1,7a	|1.6|	
X = OTs, Y = H$_{6x}$	1,2x	|4.8|	356
	1,6x	|5.0|	
	3n,4	|1.1|	
	4,5x	|4.0|	
	5x,6x	|10.3|	
	1,7s	|1.8|	
	1,7a	|1.8|	
X = I, Y = (CH$_3$)$_{6x}$	1,2x	|5.1|	356
	3n,4	|1.0|	
	4,5x	|4.7|	
	4,5n	|0.7|	
	1,7s	|1.4|	
	4,7s	|1.6|	
	1,7a	|1.6|	
	4,7a	|1.6|	
X = Br, Y = (CH$_3$)$_{6x}$	1,2x	|5.0|	356
	3n,4	|0.6|	
	4,5x	|4.1|	
	4,5n	|0.8|	
	1,7s	|1.4|	
	4,7s	|1.6|	
	1,7a	|1.3|	
	4,7a	|1.3|	
X = Br, Y = H$_{6x}$	1,2x	|5.0|	356
	1,6x	|4.9|	
	3n,4	|0.5|	
	4,5x	|3.8|	
	4,5n	|0.8|	
	5n,6x	|2.0|	
	5x,6x	|10.6|	
	1,7s	|1.4|	
	4,7s	|1.4|	
	1,7a	|1.5|	
	4,7a	|1.5|	

TABLE 24 (*continued*)

TABLE 24 *(continued)*

Compound	Proton Positions	$^3J_{HH}$ (Hz)^a	References
X = OTs	4,5x	|4.7|	356
	5x,6x	|10.8|	
	5n,6x	|2.6|	
X = OAc	1,2x	|5.2|	356
	1,6x	|4.0|	
	3n,4	|1.1|	
	5x,6x	|10.0|	
	5n,6x	|2.2|	
	1,7s	|1.7|	
	4,7s	|1.7|	
	1,7a	|1.4|	
	4,7a	|1.4|	

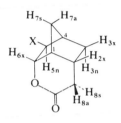

Compound	Proton Positions	$^3J_{HH}$ (Hz)^a	References
X = I	1,2x	ca. |4.|	361
	1,6x	|5.0|	
	1,7s	|1.8|	
	1,7a	|1.6|	
	2x,3n	|4.0|	
	2x,3x	|12.0|	
	3x,4	|5.0|	
	4,7s	|1.8|	
	4,7a	|1.6|	
	5n,6x	|1.9|	
X = Br	1,2x	ca. |4.|	361
	1,6x	|4.4|	
	1,7s	|1.8|	
	1,7a	|1.6|	
	2x,3n	|3.6|	
	2x,3x	|12.0|	
	3x,4	|5.0|	
	4,7s	|1.8|	
	4,7a	|1.6|	
	5n,6x	|1.6|	
X = HgOAc	1,2x	|4.3|	
	1,6x	|4.4|	
	1,7a	|1.8|	
	2x,3n	|4.2|	
	2x,3x	|12.0|	

TABLE 24 *(continued)*

TABLE 24 (*continued*)

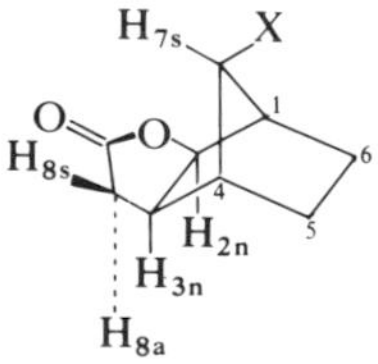

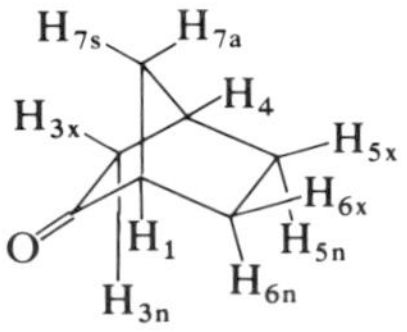

Compound	Proton Positions	$^3J_{HH}$ (Hz)^a	References
X = HgBr	3x,4	|4.5|	361
	4,7a	|1.8|	
	5n,6x	|2.8|	
	1,2x	|5.8|	
	1,6x	|5.0|	
	2x,3n	|4.0|	
	2x,3x	|12.0|	
	3x,4	|4.5|	
	5n,6x	|3.0|	

X = H$_{7a}$	1,6x	ca. |4.|	361
	2n,3n	|6.5|	
	3n,8a	|11.1|	
	3n,8s	|3.5|	
X = Br	1,7s	|1.5|	361
	2n,3n	|7.5|	
	3n,8a	|11.3|	
	3n,8s	|3.5|	
	4,7s	|1.5|	

	1,7s	+1.64	342
	1,7a	+1.19	
	4,7s	+1.60	
	4,7a	+2.10	
	1,6x	+4.72	
	1,6n	+0.12	
	3x,4	+4.76	
	3n,4	0.0	
	4,5x	+4.30	
	4,5n	+0.12	
	5x,6x	+12.05	
	5n,6n	+9.12	
	5n,6x	+4.70	
	5x,6n	+4.59	

TABLE 24 (*continued*)

TABLE 24 (*continued*)

Compound	Proton Positions	$^3J_{HH}$ (Hz)[a]	References

Compound	Proton Positions	$^3J_{HH}$ (Hz)[a]	References
W = S—⟨C₆H₄⟩—CH₃, X = H₆ₓ,	1,7a	|2.0|	360
	1,7s	|2.0|	
Y = H₅ₙ, Z = Cl	4,7a	|2.0|	
	4,7s	|2.0|	
	1,6x	|4.0|	
	5n,6x	|3.5|	
W = CCl₃, X = H₆ₓ,	1,7a	|1.5|	360
Y = H₅ₙ, Z = Br	1,7s	|1.5|	
	4,7a	|1.5|	
	4,7s	|1.5|	
	1,6x	|3.5|	
	5n,6x	|5.0|	
W = CCl₃, X = H₆ₓ,	1,7a	|1.5|	360
Y = H₅ₙ, Z = Cl	1,7s	|1.5|	
	4,7a	|1.5|	
	4,7s	|1.5|	
	1,6x	|4.0|	
	5n,6x	|5.0|	
W = SiCl₃, X = D,	1,7a	|1.5|	360
Y = H₅ₙ, Z = H₆ₙ	1,7s	|1.5|	
	4,7a	|1.5|	
	4,7s	|1.5|	
	5n,6n	|9.0|	
W = Y = D, Y = H₅ₙ,	1,7a	|2.0|	360
Z = H₆ₙ	1,7s	|2.0|	
	4,7a	|2.0|	
	4,7s	|2.0|	
	5n,6n	|6.8|	

Compound	Proton Positions	$^3J_{HH}$ (Hz)[a]	References
X = H₂ₓ, Y = OH	2x,3x	|9.6|	425
	2x,3n	|3.5|	

TABLE 24 (*continued*)

TABLE 24 (*continued*)

Compound	Proton Positions	$^3J_{HH}$ (Hz)[a]	References
X = OH, Y = H$_{2n}$	2n,3x	|5.2|	425
	2n,3n	|5.2|	
X = H$_{2x}$, Y = OAc	2x,3x	|9.8|	425
	2x,3n	|3.|	
X = OAc, Y = H$_{2n}$	2n,3x	|5.1|	425
	2n,3n	|5.1|	
X = H$_{2x}$, Y = HO—◯— 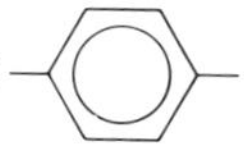	2x,3x	|11.4|	425
	2x,3n	|5.5|	
X = HO—◯—,  Y = H$_{2n}$	2n,3x	|8.3|	425
	2n,3n	|8.3|	
X = H$_{2x}$, Y = CO$_2$H	2x,3x	|8.0|	425
	2x,3n	|8.0|	
X = H$_{2x}$, Y = Cl	2x,3x	|9.9|	425
	2x,3n	|4.2|	
X = Cl, Y = H$_{2n}$	2n,3x	|8.0|	425
	2n,3n	|5.4|	
X = H$_{2x}$, Y = OPh	2x,3x	|9.3|	425
	2x,3n	|2.8|	
X = OPh, Y = H$_{2n}$	2n,3x	|5.2|	425
	2n,3n	|5.2|	
X = H$_{2x}$, Y = MeO—◯— 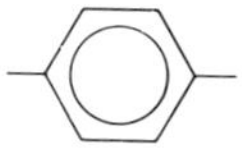	2x,3x	|11.4|	425
	2x,3n	|5.8|	
X = MeO—◯—, 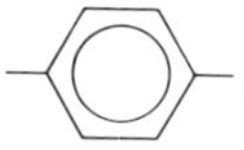 Y = H$_{2n}$	2n,3x	|8.7|	425
	2n,3n	|8.7|	
X = H$_{2x}$, Y = HO—◯ 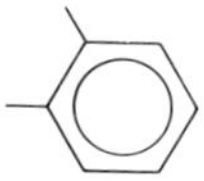	2x,3x	|11.5|	425
	2x,3n	|6.0|	
X = HO—◯, 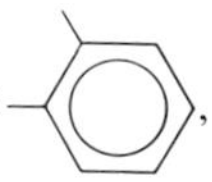 Y = H$_{2n}$	2n,3x	|8.7|	425
	2n,3n	|8.7|	
X = H$_{2x}$, Y = MeO—◯ 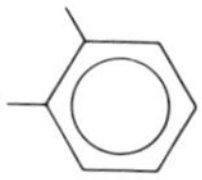	2x,3n	|6.0|	425

TABLE 24 (*continued*)

TABLE 24 (*continued*)

Compound	Proton Positions	$^3J_{HH}$ (Hz)[a]	References
X = MeO—⟨benzene ring⟩, Y = H$_{2n}$ 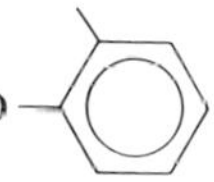	2n,3x	|8.8|	425
	2n,3n	|8.9|	
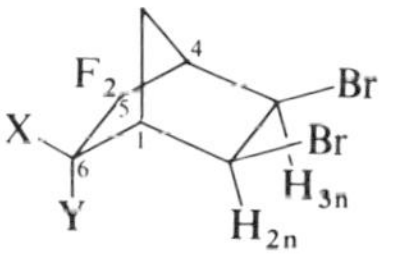 *(F$_2$ bicyclic structure: positions 4,5,6,1; substituents X, Y, Br, Br, H$_{3n}$, H$_{2n}$)*			
X = Y = F	1,6x	ca. |5.|	363
X = Y = CF$_3$	2n,3n	|7.0|	
X = Y = Cl	2n,3n	|7.0|	
X = CF$_2$Cl, Y = Cl	2n,3n	|6.9|	
X = F, Y = CF$_2$Cl	2n,3n	|6.9|	
X = H$_{6x}$, Y = F$_{6n}$	2n,3n	|7.0|	
X = F$_{6x}$, Y = H$_{6n}$	2n,3n	|6.9|	
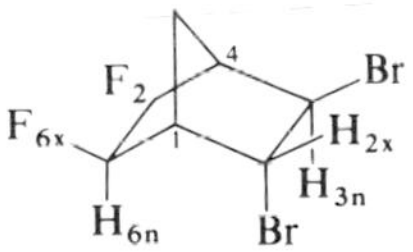 *(F$_2$ bicyclic structure: F$_{6x}$, Br, H$_{2x}$, H$_{3n}$, H$_{6n}$, Br)*	2x,3n	ca. |3.2|	363
	1,2x	ca. |3.2|	
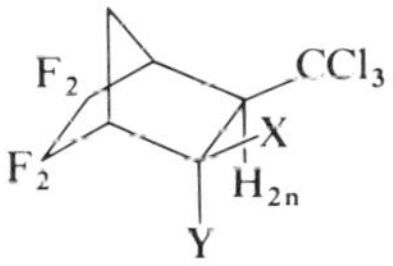 *(F$_2$, F$_2$ bicyclic with CCl$_3$, X, H$_{2n}$, Y)*			
X = Cl, Y = H$_{3n}$	2n,3n	|6.9|	364
X = Br, Y = H$_{3n}$	2n,3n	|7.1|	364
X = H$_{3x}$, Y = Cl	2n,3n	|6.9|	364
	3x,4	|3.7|	
X = H$_{3x}$, Y = Br	2n,3n	|6.7|	364
	3x,4	ca. |4.|	
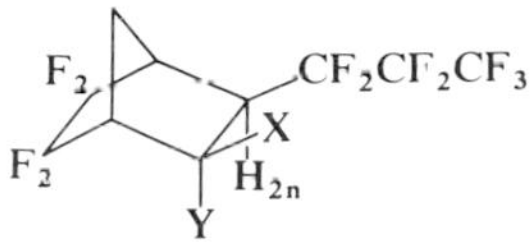 *(F$_2$, F$_2$ bicyclic with CF$_2$CF$_2$CF$_3$, X, H$_{2n}$, Y)*			
X = I, Y = H$_{3n}$	2n,3n	|7.8|	364
X = H$_{3x}$, Y = I	2n,3x	|7.8|	364

TABLE 24 (*continued*)

TABLE 24 (*continued*)

Compound	Proton Positions	$^3J_{HH}$ (Hz)^a	References
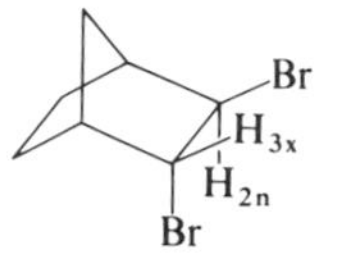 (norbornane structure) Br, H_{3x}, H_{2n}, Br	2n,3x	|2.9|	329
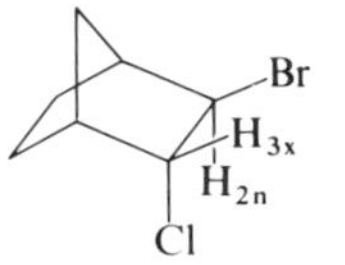 (norbornane structure) Br, H_{3x}, H_{2n}, Cl	2n,3x	|2.6|	329
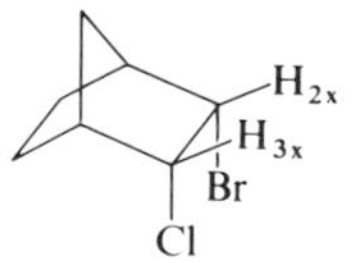 (norbornane structure) H_{2x}, H_{3x}, Br, Cl	2x,3x	|9.5|	329
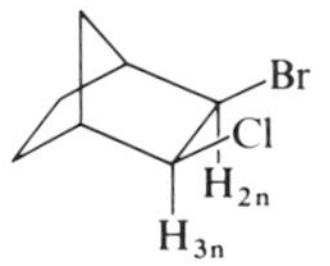 (norbornane structure) Br, Cl, H_{2n}, H_{3n}	2n,3n	|6.8|	329
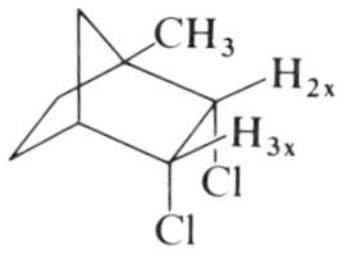 (norbornane structure) CH_3, H_{2x}, H_{3x}, Cl, Cl	2x,3x	|9.3|	329

(norbornane structure) X, Y, Cl, D_2, H_{5x}, H_{6x}, D_2, Cl, H_{5n}, H_{6n}

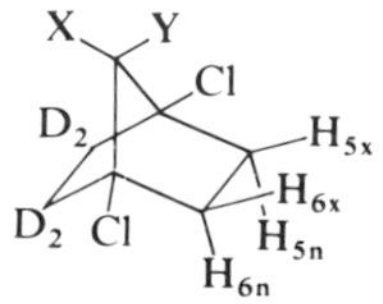 Compound	Proton Positions	$^3J_{HH}$ (Hz)^a	References
$X = H_{7s}$, $Y = H_{7a}$	5x,6x	$+13.20 \pm 0.04$	341
	5n,6n	$+9.15 \pm 0.02$	
	5x,6n = 5n,6x	$+4.74 \pm 0.03$	
$X = Y = Cl$	5x,6x	$+12.50 \pm 0.04$	341
	5n,6n	$+9.97 \pm 0.04$	
	5x,6n = 5n,6x	$+4.47 \pm 0.04$	
$X = Y = OMe$	5x,6x	$+12.71 \pm 0.02$	341
	5n,6n	$+9.82 \pm 0.02$	
	5x,6n = 5n,6x	$+4.46 \pm 0.02$	

TABLE 24 (*continued*)

TABLE 24 (*continued*)

Compound	Proton Positions	$^3J_{HH}$ (Hz)[a]	References
$X = H_{7s}$, $Y = OAc$	5x,6x	$+12.85 \pm 0.02$	341
	5n,6n	$+9.15 \pm 0.04$	
	5x,6n = 5n,6x	$+4.16 \pm 0.01$	
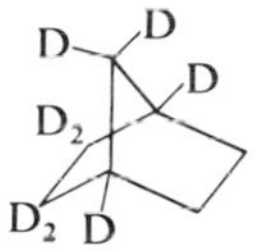	5x,6x	$+12.22$	335
	5n,6n	$+9.05$	
	5x,6n = 5n,6x	$+4.62$	
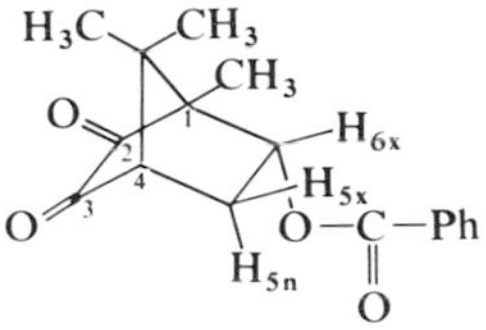	4,5n	ca. 0	359
	4,5x	$+5.6$	
	5n,6x	$+3.4$	
	5x,6x	$+9.2$	

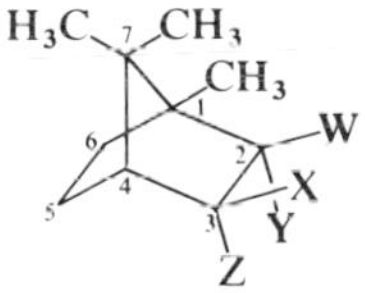 Compound	Proton Positions	$^3J_{HH}$ (Hz)[a]	References
$W = X = OH$, $Y = H_{2n}$, $Z = H_{3n}$	2n,3n	$|7.7|$	426
	3n,4	0	
$W = Z = OH$, $X = H_{3x}$, $Y = H_{2n}$	2n,3x	$|2.3|$	426
	3x,4	ca. $|4.0|$	
$W = H_{2x}$, $X = H_{3x}$, $Y = Z = OH$	2x,3x	$|8.9|$	426
	3x,4	$|4.4|$	
$W = H_{2x}$, $X = Y = OH$, $Z = H_{3n}$	2x,3n	$|2.2|$	426
	3n,4	0	
$W = H_{2x}$, $X = H_{3x}$, $Y = OH$, $Z = NMe_3^+$ Br^-	2x,3x	$|9.8|$	426
	3x,4	$|4.0|$	
$W = OH$, $X = H_{3x}$, $Y = H_{2n}$, $Z = NMe_3^+$ Br^-	2n,3x	$|5.3|$	426
	3x,4	$|3.5|$	

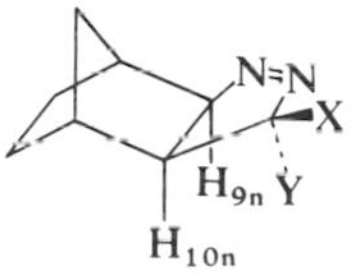 Compound	Proton Positions	$^3J_{HH}$ (Hz)[a]	References
$X = Ph$, $Y = \overset{O}{\overset{\|}{P}}(OMe)_2$	9n,10n	$|7.0|$	345
$X = Me$, $Y = \overset{O}{\overset{\|}{P}}(OMe)_2$	9n,10n	$|7.1|$	345

TABLE 24 (*continued*)

TABLE 24 (*continued*)

Compound	Proton Positions	$^3J_{HH}$ (Hz)^a	References
$X = P(OMe)_2$ (with O double bond), $Y = Me$	9n,10n	|6.0|	345
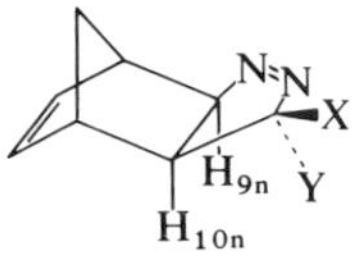			
$X = Ph$, $Y = P(OMe)_2$ (with O double bond)	9n,10n	|6.4|	345
$X = Me$, $Y = P(OMe)_2$ (with O double bond)	9n,10n	|6.5|	345
$X = P(OMe)_2$ (with O double bond), $Y = Me$	9n,10n	|6.1|	345
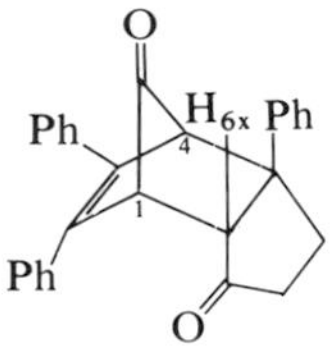	1,6x	|5.0|	427
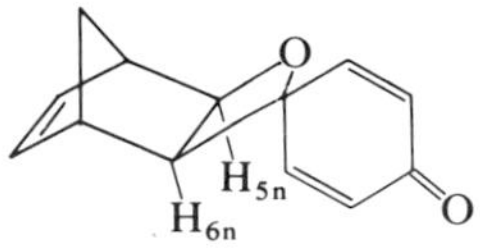	5n,6n	|5.|	366
	5n,6n	|6.|	366
	1,6x	|5.|	
	1,7s = 4,7s	|2.|	
	3x,4	|3.|	366
	2,3x	ca. |1.|	

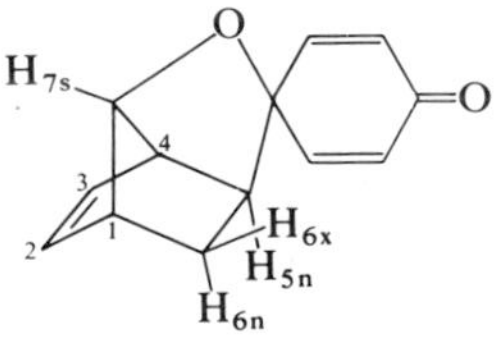

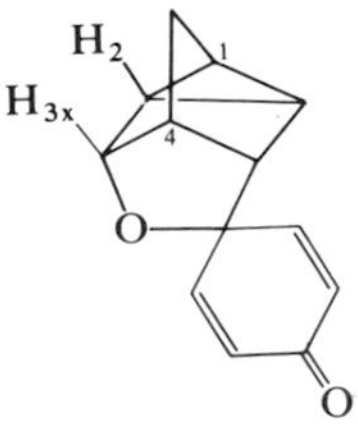

TABLE 24 (*continued*)

TABLE 24 (*continued*)

Compound	Proton Positions	$^3J_{HH}$ (Hz)[a]	References
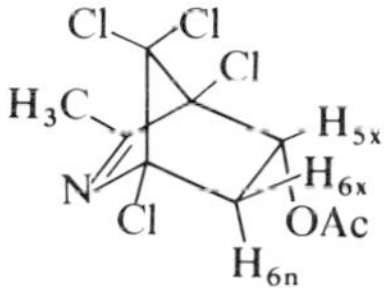	5x,6n 5x,6x	|2.4| |7.8|	371
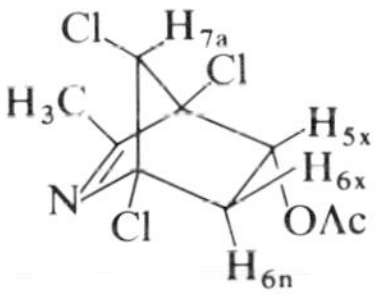	5x,6n 5x,6x	|2.9| |8.3|	371
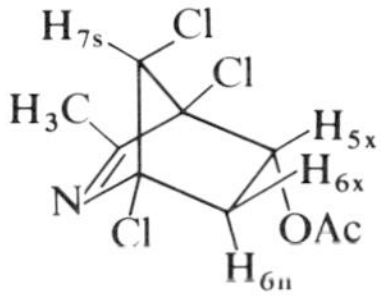	5x,6n 5x,6x	|2.7| |8.1|	371
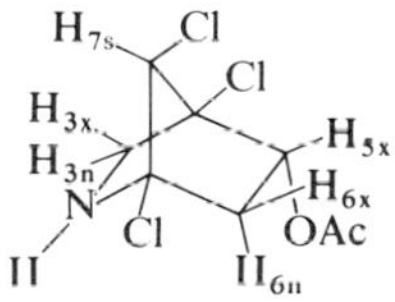	5x,6n 5x,6x	|2.8| |10.0|	371
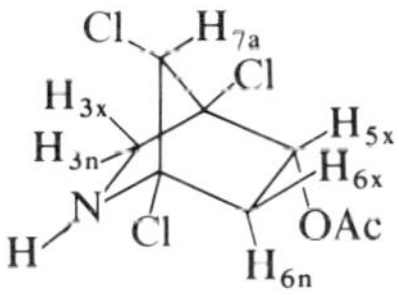	5x,6n 5x,6x	|3.7| |10.3|	371
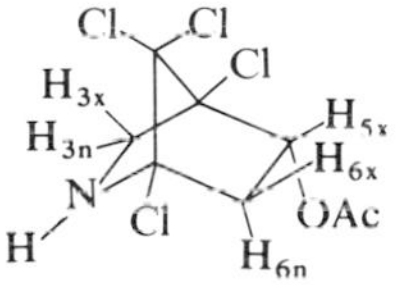	5x,5n 5x,6x	ca. |3.| |10.0|	371
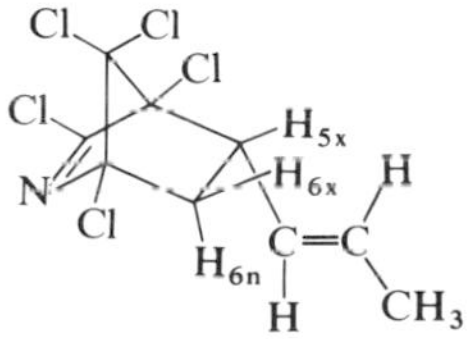	5x,6n 5x,6x	+3.88 +8.47	368

TABLE 24 (*continued*)

TABLE 24 (*continued*)

Compound	Proton Positions	$^3J_{HH}$ (Hz)[a]	References

The compound structure (bicyclic with Cl substituents and positions H_{6x}, H_{5x}, H_{6n}, X):

X = (C=C with H, H / H, CH₃)	5x,6x 5x,6n	|8.5| |4.0|	369
X = (C=C with H, CH₃ / H, H)	5x,6x 5x,6n	|8.5| |4.0|	369
X = (C=C with H₃C, H / H, CH₃)	5x,6x 5x,6n	|8.5| |4.0|	369
X = (C=C with H₃C, H / CH₃, H)	5x,6x 5x,6n	|8.5| |4.0|	369
X = (C=C with H, CH₃ / CH₃, CH₃)	5x,6x 5x,6n	|9.5| |4.0|	369
(bicyclic with N–CH₃, positions H_{6x}, H_{5x}, H_{6n}, H_{5n})	5x,6x 5n,6n 5n,6x	|10.0| |8.6| |4.0|	370

$X = \overset{O}{\overset{\|}{C}}CH_3$, $Y = CO_2Mr$

$X = \overset{O}{\overset{\|}{C}}CH_3$, $Y = CO_2Mr$	2n,3x	|5.|	428

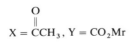

TABLE 24 (*continued*)

TABLE 24 (*continued*)

Compound	Proton Positions	$^3J_{HH}$ (Hz)^a	References
X = Y = CO$_2$Me	2n,3x	|5.|	428
X = Y = CO$_2$H	2n,3x	|5.|	428
	3x,4	|5.|	

| 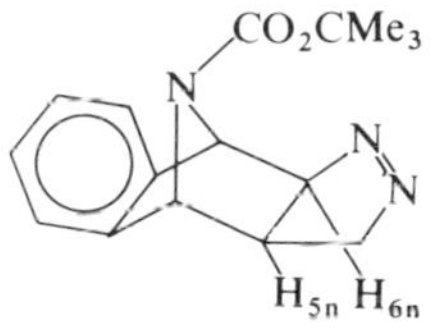 | 5n,6n | \|7.4\| | 429 |

| 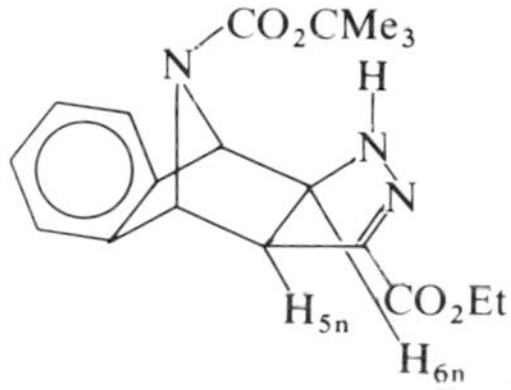 | 5n,6n | \|9.0\| | 429 |

| 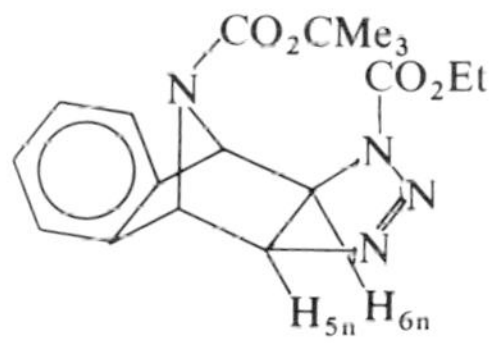 | 5n,6n | \|8.0\| | 429 |

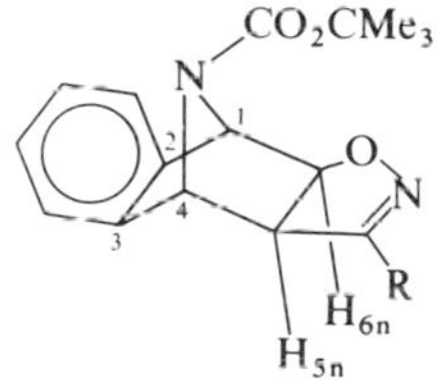 R = PhC=O	1,6n	0	429
	4,5n	0	
	5n,6n	|7.5|	
R = 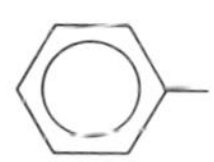	1,6n	0	429
	4,5n	0	
	5n,6n	|7.6|	
R = Cl— 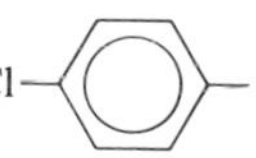	1,6n	0	429
	4,5n	0	
	5n,6n	|7.8|	

TABLE 24 (*continued*)

TABLE 24 (*continued*)

Compound	Proton Positions	$^3J_{HH}$ (Hz)^a	References
 R = MeO$_2$C—⟨⟩—	1,6n 4,5n 5n,6n	0 0 |7.5|	429
 R = O$_2$N—⟨⟩—	1,6n 4,5n 5n,6n	0 0 |7.5|	429

| R = Ph | 1,2 = 7,8
2,3 = 6,7 | 0
\|2.2\| | 430 |
| R = Me | 1,2 = 7,8
2,3 = 6,7 | 0
\|3.0\| | 430 |

| 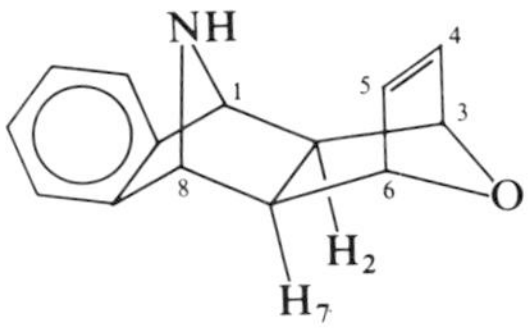 | 1,2 = 7,8
2,3 = 6,7 | 0
\|2.3\| | 430 |

| 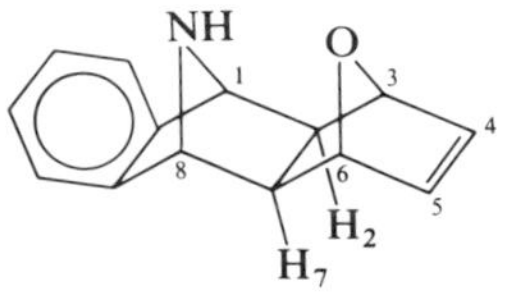 | 1,2 = 7,8
2,3 = 6,7 | 0
0 | 430 |

| 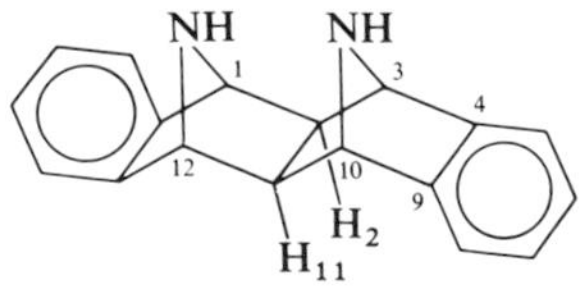 | 1,2 = 11,12
2,3 = 10,11 | 0
0 | 430 |

| 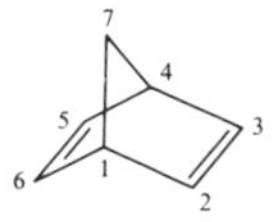 | 1,2
2,3
1,7 | \|2.9 ± 0.1\|
\|3.45 ± 0.1\|
0 | 431 |

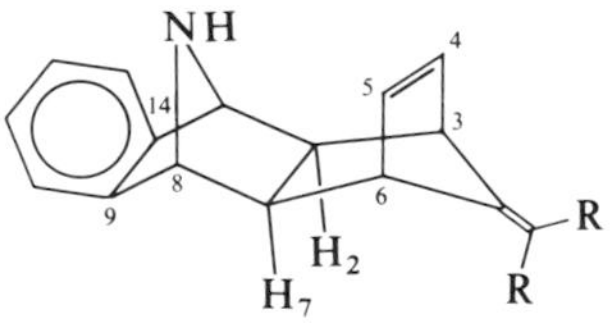

TABLE 24 (*continued*)

TABLE 24 (*continued*)

Compound	Proton Positions	$^3J_{HH}$ (Hz)[a]	References
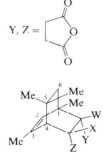	1,2	|6.1|	220
	1,6	|1.8|	
	1,7	|2.8|	
	2,3	|4.6|	
	1,7	|1.3|	124

$W = X = Y = Z = H$	"1,7"	|1.6|	124
	2,3	|8.24|	267
$W = H_{5x}, X = H_{6x},$	5x,6x	|8.3|	124
$Y = OH, Z = H_{6n}$	5x,6n	|2.6|	
	"4,5"	|2.6|	
$W = H_{5x}, X = H_{6x},$	5x,6x	|8.4|	124
$Y = OAc, Z = H_{6n}$	5x,6n	|2.6|	
	"4,5"	|2.6|	
$W = H_{5x}, X = H_{6x},$	1,6x	ca. |0.8|	124
$Y = Z = CO_2Me$			
$W = Z = CO_2Me, X = H_{6x},$	4,5x	|1.8|	124
$Y = H_{5n}$	5n,6x	|6.0|	
$W = H_{5x}, X = H_{6x},$	1,6x	|1.4|	124

$Y, Z =$

$W = H_{7x}, X = H_{8x},$	4,8x	|3.5|	135

TABLE 24 (*continued*)

TABLE 24 (*continued*)

Compound	Proton Positions	$^3J_{HH}$ (Hz)^a	References
Y, Z = 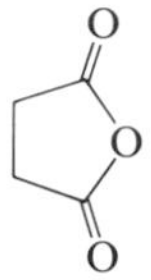	7x,8x	|8.5|	
W = H_{7x}, X = H_{8x},	4,8x	|2.0|	135
Y = Z = CO_2Me	7x,8x	|10.9|	
W = Z = CO_2Me,	4,8x	|2.0|	135
X = H_{8x}, Y = H_{7n}	7n,8x	|5.9|	
W = H_{7x}, X = Y = CO_2Me,	4,8n	|1.9|	135
Z = H_{8n}	7x,8n	|8.5|	
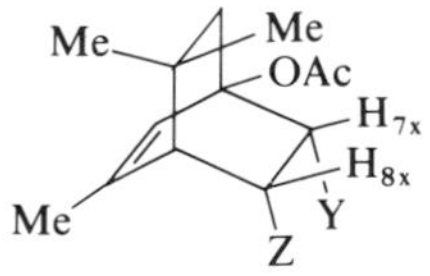			
X = Y = CO_2Me	7x,8x	|12.0|	352
X,Y = 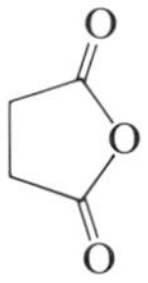	7x,8x	|10.0|	352
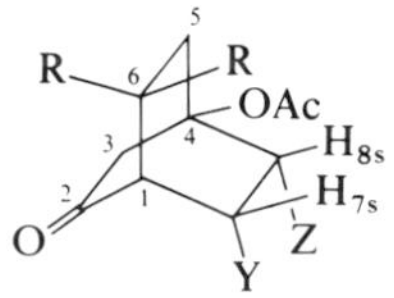			
R = H_6, Y = Z = CO_2Me	7s,8s	|12.0|	352
R = Me, Y = Z = CO_2Me	7s,8s	|12.0|	352
R = Me, Y, Z = 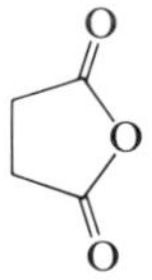	7s,8s	|9.5|	352
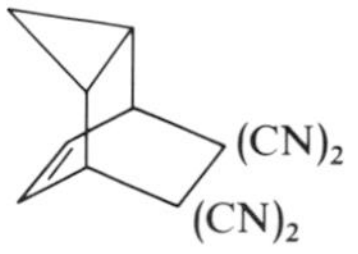	1,2	|6.91|	432
	2,3	|8.12|	432

TABLE 24 (*continued*)

TABLE 24 (*continued*)

Compound	Proton Positions	$^3J_{HH}$ (Hz)[a]	References

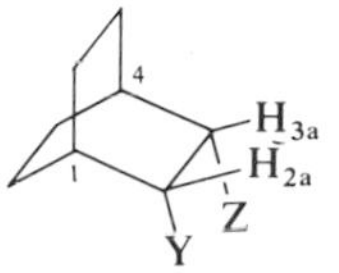			
Y, Z 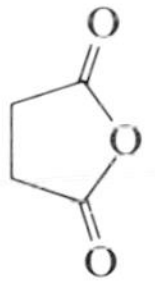 =	1,2	|1.6|	124
Y = OH, Z = H$_{3s}$	1,2	|2.8|	124
	2a,3a	|9.0|	
	2a,3s	|2.8|	
Y = OAc, Z = H$_{3s}$	1,2	|2.8|	124
	2a,3a	|9.0|	
	2a,3s	|2.8|	

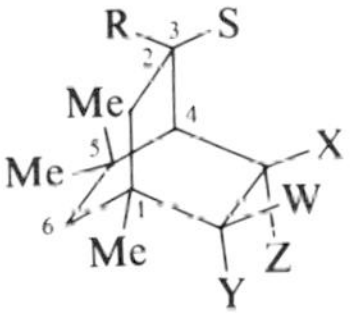

R = H$_{3a}$, S = Me, Y = H$_{7a}$,	3a,4	|2.0|	135
	3a,(Me)$_{3s}$	|7.5|	
	4,8a	|3.5|	
	7a,8a	|10.8|	
Z = H$_{8a}$, W, X =			
R = H$_{3a}$, S = Me,	3a,4	|2.2|	135
W = X = CO$_2$Me,	3a,(Me)$_{3s}$	|7.2|	
Y = H$_{7a}$, Z = H$_{8a}$	4,8a	|2.3|	
	7a,8a	|11.9|	
R = Me, S = H$_{3s}$, W = H$_{7s}$,	3s,(Me)$_{3a}$	|6.8|	135
X = Y = CO$_2$Me,	4,8a	|1.2|	
Z = H$_{8a}$	7s,8a	|8.3|	
R = H$_{3a}$, S = Me,	3a,4	|1.5|	135
W = Z = CO$_2$Me,	3a,(Me)$_{3s}$	|6.7|	
X = H$_{8s}$, Y = H$_{7a}$	4,8s	|1.5|	
	7a,8s	|9.9|	

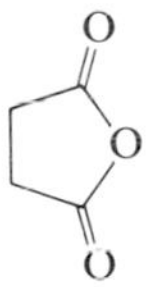

TABLE 24 (*continued*)

TABLE 24 (*continued*)

Compound	Proton Positions	$^3J_{HH}$ (Hz)[a]	References
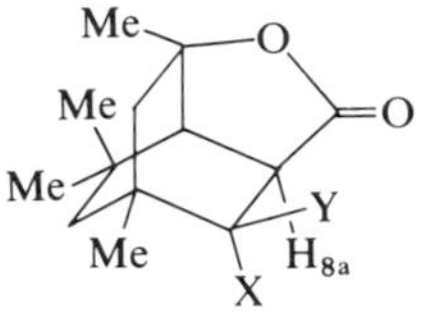 $X = H_{7a}$, $Y = CO_2Me$	4,8a	|4.4|	135
	7a,8a	|9.7|	
$X = CO_2Me$, $Y = H_{7s}$	4,8a	|4.3|	135
	7s,8a	|2.1|	
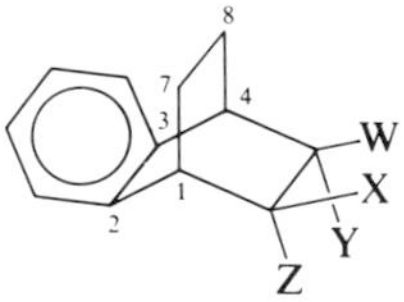 $W = H_{5x}$, $X = OH$,	1,6n	|2.7|	124
$Y = H_{5n}$, $Z = H_{6n}$	5n,6n	|9.2|	
	5x,6n	|3.2|	
$W = H_{5x}$, $X = H_{6x}$,	1,6x	|2.7|	124
$Y = H_{5n}$, $Z = OH$	4,5x	|2.7|	
	5x,6x	|8.8|	
	5n,6x	|3.2|	
$W = H_{5x}$, $X = OAc$,	1,6n	|2.7|	124
$Y = H_{5n}$, $Z = H_{6n}$	5n,6n	|9.2|	
	5x,6n	|3.2|	
$W = H_{5x}$, $X = H_{6x}$,	1,6x	|2.7|	124
$Y = H_{5n}$, $Z = OAc$	4,5x	|2.7|	
	5x,6x	|9.2|	
	5n,6x	|3.2|	
$W = H_{5x}$, $X = H_{6x}$,	1,6x	|1.8|	124
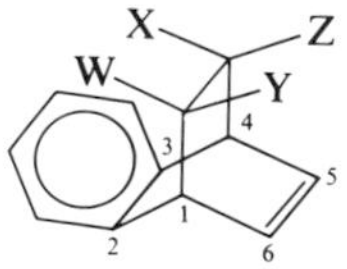 $Y = Z = CO_2Me$			
$W = H_{7s}$, $X = H_{8s}$,	"1,7"	|1.4|	124
$Y = H_{7a}$, $Z = H_{8a}$	5,6	|7.6 ± 0.1|	267
$W = H_{7s}$, $X = H_{8s}$,	1,7s	|0.9|	124
$Y = Z = CO_2Me$			
$W = Z = CO_2Me$,	4,8s	|2.2|	124
$X = H_{8s}$, $Y = H_{7a}$	7a,8s	|5.0|	

TABLE 24 (*continued*)

TABLE 24 (*continued*)

Compound	Proton Positions	$^3J_{HH}$ (Hz)^a	References
W = X − CO$_2$Me, Y = H$_{7a}$, Z = H$_{8a}$	1,7a	|1.0|	124
X = H$_7$	1,7	|1.3|	124
X = NH$_2$	1,7	|2.5|	353
	4,8	|2.5|	
	7,8s	|3.0|	
	7,8a	|9.5|	
X = Cl	1,7	|2.5|	353
	4,8	|2.5|	
	7,8s	|2.5|	
	7,8a	|8.0|	
X = OH	1,7	|3.0|	353
	4,8	|3.0|	
	7,8s	|3.0|	
	7,8a	|8.5|	
X = OAc	1,7	|2.6|	353
	4,8	|2.6|	
	7,8s	|3.3|	
	7,8a	|8.8|	
X = OTs	1,7	|3.0|	353
	4,8	|2.5|	
	7,8s	|3.5|	
	7,8a	|9.0|	
X = SPh	1,7	|2.8|	353
	4,8	|2.6|	
	7,8s	|4.6|	
	7,8a	|9.2|	
X = $\overset{+}{N}$Me$_3$ Br$^-$	1,7	|2.0|	405
	4,8	|2.7|	
	7,8s	|5.8|	
	7,8a	|10.0|	

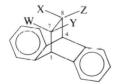

TABLE 24 (*continued*)

TABLE 24 (*continued*)

Compound	Proton Positions	$^3J_{HH}$ (Hz)^a	References
W = H$_7$, X = SPh, Y = Cl, Z = H$_8$	1,7	|3.0|	353
	4,8	|2.5|	
	7,8 (*trans*)	|3.0|	
W = H$_7$, X = H$_8$, Y = Cl, Z = SPh	1,7	|2.0|	353
	4,8	|2.0|	
	7,8 (*cis*)	|9.0|	
W = H$_7$, X = SMe, Y = Cl, Z = H$_8$	1,7	|2.7|	353
	4,8	|2.4|	
	7,8 (*trans*)	|3.6|	
W = H$_7$, X = H$_8$, Y = Cl, Z = SCF$_3$	1,7	|2.3|	353
	4,8	|2.3|	
	7,8 (*cis*)	|8.2|	
W = H$_7$, X = H$_8$, Y = Cl,	1,7	|3.0|	353
	4,8	|2.0|	
Z = S-⟨◯⟩-Me	7,8 (*cis*)	|8.6|	
W = H$_7$, X = OAc, Y = Cl, Z = H$_8$	1,7	|2.5|	353
	4,8	|2.5|	
	7,8 (*trans*)	|2.5|	
W = H$_7$, X = H$_8$, Y = Cl, Z = OAc	4,8	|2.5|	353
	7,8 (*cis*)	|8.0|	
W = H$_7$, X = OAc, Y = I, Z = H$_8$	1,7	|3.0|	353
	4,8	|3.0|	
	7,8 (*trans*)	|3.0|	
	4,8	|2.6|	353

^{a} The absolute sign of $^3J_{HH}$ couplings across sp^3-hybridized carbon atoms is generally considered to be positive.[421]

suggest that "interactions of the C(7)methylene bridge with the bonds of the C(2)—C(3) ethylene bridge are responsible for the nonequivalence of $J_{exo,exo}$ and $J_{endo,endo}$ in norbornanes, and that in norbornenes interaction of the olefin functionality with bonds of the ethylene bridge is responsible for bridging $J_{exo,exo}$ back fortuitously close to $J_{endo,endo}$."[335]

In addition to the factors discussed above, the fact that the magnitudes of vicinal proton–proton coupling constants are sensitive to electronegativity effects of substituents is well documented.[435,436] Recently, an attempt has been made to modify the $^3J_{HH}$-torsion angle (Karplus) relation to take into account electronegativity effects of substituents and the orientation of each substituent relative to the coupled protons in substituted ethanes.[436] Interestingly, $^3J_{HH}$ values in norbornanes and norbornenes were excluded from the data set that was used to determine empirically the parameters in this modified Karplus relation. This decision rested upon two considerations: (i) the fact that the presence of substituents is known to result in considerable twisting (i.e., distortion) of the norbornane skeleton,[433] with consequent reduction in the degree of certainty with which the geometry of the bicyclic system can be defined, and (ii) the existence of an "extra" coupling pathway involving inter- action of the methylene bridge with the C(2)—C(3) fragment[335] (discussed above)

Some vicinal carbon–hydrogen couplings measured in rigid bicyclic systems appear in Table 25. These couplings have been the subject of recent extensive theoretical studies by Barfield and co-workers.[383,411,438,439] In particular, these investigators have found that the familiar γ effect on ^{13}C nuclear shield- ings (cf. Steric Influences of Substituents on NMR Chemical Shifts, Carbon-13 Chemical Shifts, Chapter 3) has its counterpart in $^3J_{CX}$ couplings, i.e., the presence of γ-methyl or γ-methylene substituents (R) on C(3) in **124** has been

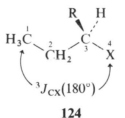

124

shown (INDO-FPT calculations) to decrease the magnitude of $^3J_{CX}(180°)$ couplings involving C(1).[438,439] Therefore, the importance of the contribution of nonbonded interactions to $^3J_{CX}(180°)$ in these systems must be recognized and taken into account if these couplings are to be utilized properly for purposes related to conformational and configurational analysis. In a similar vein, nonadditivity of bridgehead $^{13}C—^1H$ and $^{13}C—^{13}C$ coupling constants along the series **125–128** has been ascribed[383] to the operation of nonbonded

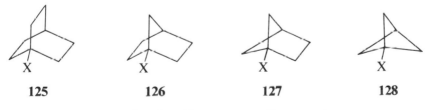

$$(X = {}^1H, {}^2H, {}^{13}CO_2H, \text{ and } {}^{13}CH_2OH)$$

TABLE 25. $^3J_{CH}$ Values in Rigid Bicyclic Ring Systems Related to Norbornane and Bicyclo[2.2.2]octane

Compound	Carbon Atom Position	Proton Position	$^3J_{CH}$ (Hz)	Reference
	3	1	|7.0|	383
	3	1	|7.0|	383
X = $^{13}CO_2H$	$^{13}C{=}O$ $^{13}C{=}O$	6n 6x	+ 9.10 + 3.70	336 336
	3 1 1	5x 6x 6n	|8.6| |9.8| |9.8|	437
	7(C=O)	5n	|7.4|	375
	7(C=O)	5x	0	375

TABLE 25 (*continued*)

TABLE 25 (continued)

Compound	Carbon Atom Position	Proton Position	$^3J_{CH}$ (Hz)	Reference
	7(C=O)	5n	|7.4|	375
	7(C=O)	5x	0	375
	7(C=O)	5n	|7.4|	375
X = H	7(C=O)	5n	|7.0|	375
X = Cl	7(C=O)	5n	|6.3|	
	7(C=O)	5x	0	375
	7(C=O)	5x	0	375
	7(C=O)	6n	|7.1|	

interactions between the bridgehead carbons. The results of INDO-FPT calculations suggest that (nonbonded) bridgehead–bridgehead interactions make increasingly negative contributions to these couplings as the distance between bridgehead carbon atoms decreases along the series $125 \rightarrow 128$ (cf. discussion of $^3J_{CC}$ coupling constants in this section).

Marshall and Seiwell[336] have measured a variety of $^3J_{CH}$ values in ^{13}C-*carboxyl*-labeled norbornene- and norbornanecarboxylic acids. Spin tickling proton NMR experiments on ^{13}C-*carboxyl*-labeled 1,2,3,4,7,7-hexachloronorborn-2-ene-*endo*-5-carboxylic acid (129) established that cis and trans vicinal carbon–hydrogen couplings (i.e., $^3J_{6n, 8}$, and $^3J_{6x, 8}$, respectively) are positive in this system.[336]

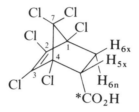

(asterisk indicates position of ^{13}C)

129

Interestingly, a good quality linear correlation between $^nJ_{HH}$ and $^nJ_{CH}$ was obtained for $n = 2, 3, 4,$ and 5 (correlation coefficient = 0.975).[336] A total of 20 J values in a variety of "geometrically equivalent model systems" were employed in preparing this correlation. This result was considered to reflect the fact that $^nJ_{CH}$ and $^nJ_{HH}$ couplings operate via similar mechanisms over the range of systems studied.[336] A similar comparison of $^nJ_{HF}$ with $^nJ_{HH}$ revealed a lack of linear correlation (correlation coefficient = -0.098), supporting the previously held view[440] that these couplings operate by quite different mechanisms. In fact, there exists a convincing body of evidence suggesting that the dihedral angle dependence of $^3J_{HF}$ is adequately described by a Karplus-type relation (see below, in this section); therefore, $^3J_{HF}$ couplings would be expected to correlate linearly with $^3J_{HH}$ in structurally similar systems. The culprits in the Marshall and Seiwell study[336] appear to be long-range $^nJ_{HF}$ couplings ($n \geq 4$); these will be further discussed under Four-Bond and More Distant (Long-Range) Couplings, below.

Vicinal carbon–proton couplings have been utilized to assign stereochemistry in a variety of Diels-Alder adducts.[336] Diels-Alder addition of substituted cyclopentadienones (e.g., 130) to substituent-bearing dienophiles (131) results in the formation of substituted 7-ketonorbornenes (e.g., 132 and/or 133):

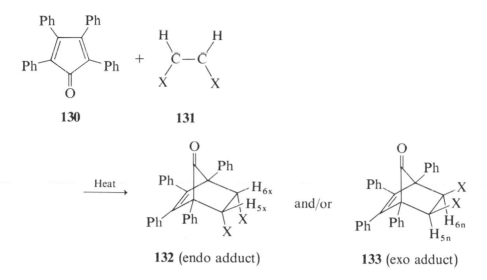

132 (endo adduct) and/or **133** (exo adduct)

Coupling between 5n,6n protons and the C(7) carbonyl group in the exo adducts (**133**) was found to be readily discernible [$^3J_{CH}$ = ca. 6.3–7.6 Hz, dihedral angle ϕ = 160–170° between C(7) and the 5n,6n protons]. However, in the endo adducts (**132**), the dihedral angle between C(7) and the 5x,6x protons is essentially zero. Therefore, observation of the multiplicity of the C(7) signal in the proton-coupled ^{13}C NMR spectra of the Diels-Alder adducts was found to be reliable for differentiating between exo and endo adducts of the type **132** and **133** and also for assigning configuration in instances where the Diels-Alder reaction (**130** + **131**) afforded only one of two possible diastereoisomeric cycloadducts.[425]

Representative vicinal $^3J_{HF}$ values measured in rigid bicyclic systems are presented in Table 26. The dihedral angle and bond angle dependence of vicinal proton–fluorine couplings in a variety of rigid bicyclic systems has been studied by Williamson and co-workers.[376,398] They find the behavior of $^3J_{HF}$ to be "almost completely analogous to $^3J_{HH}$," showing Karplus-type dihedral angle dependence similar to that displayed by $^3J_{HH}$.[376,398] Subsequently, theoretical calculations for $^3J_{HF}$ in saturated systems (FCH$_2$–CH$_2$F and FCH$_2$–CF$_3$) revealed clear, Karplus-type dihedral angle dependence (i.e., calculated $^3J_{HF}$ values displayed maximum values at 0° and 180° and minimum values near 90° and 270°, the $^3J_{HF}$ values at ϕ = 180° being somewhat greater than those at ϕ = 0°).[235,441]

Williamson et al.[398] recognized that the magnitude of $^3J_{HF}$ probably depended on other factors in addition to dihedral angle, specifically (i) on the electronegativity of adjacent substituents, and (ii) on the length of the C—H, C—F, and C—C bonds that intervene along the three-bond pathway which connects the coupled fluorine and hydrogen nuclei. The importance of substituent electronegativity in this regard was verified by Ihrig and Smith[377] in

TABLE 26. $^3J_{FH}$ Values in Rigid Bicyclic Ring Systems Related to Norbornane and Bicyclo[2.2.2]octane

Compound	Fluorine Atom Position	Proton Position	$^3J_{FH}$ (Hz)	Reference
$W = H_{5x}$, $X = F_{6x}$,	5n	6n	$\|17.71\|$	377
$Y = F_{5n}$, $Z = H_{6n}$	6x	5x	$\|13.85\|$	
$W = H_{5x}$, $X = H_{6x}$,	5n	6x	$\|1.87\|$	377
$Y = F_{5n}$, $Z = F_{6n}$				
$W = H_{5x}$, $X = F_{6x}$,	6n	5x	$\|0.31\|$	377
$Y = F_{5n}$, $Z = F_{6n}$	6x	5x	$\|9.13\|$	
$X = I$	5x	4	$\|7.9\|$	378
	5n	4	$\|2.0\|$	
$X = F$	5x	4	$\|9.1\|$	378
	5n	4	$\|2.1\|$	
$X = CO_2H$	5x	4	$\|8.7\|$	378
	5n	4	$\|2.2\|$	
	1	7x	$\|2.3\|$	358
$X = H_{6x}$, $Y = F_{6n}$	5x	6x	$\|17.\|$	363
$X = H_{6n}$, $Y = F_{6x}$	6x	1	$\|5.7\|$	363
	5n	6n	ca. $\|9.5\|$	

TABLE 26 (*continued*)

TABLE 26 (continued)

Compound	Fluorine Atom Position	Proton Position	$^3J_{FH}$ (Hz)	Reference
X = H$_{6x}$, Y = F$_{6n}$	5x	6x	|19.5|	363
	5x	4	ca. |6.5|	
X = F$_{6x}$, Y = H$_{6n}$	5n	6n	|10.6|	363
	5n	6n	|8.3|	363
	5	6n	|24.7|	376, 398
	5	6n	|25.10|	377
	5	6x	|12.2|	376, 398
	5	6x	|12.01|	377
	5	6n	|19.8|	376, 398
	5	4	ca. |2.0|	
	5	6n	|10.55|	376, 398
	5	4	|2.3|	
	7	8	|30.8|	376, 398
	7	1	|3.8|	
	7	8	|22.5|	376, 398
	7	1	|6.0|	

their study of $^3J_{HF}$ values in a series of substituted norbornenes synthesized via Diels-Alder addition of hexachlorocyclopentadiene to several fluoroethylenes. In particular, the trans $^3J_{HF}$ coupling ($^3J_{5n,6x}$ or $^3J_{5x,6n}$) exhibited a 40-fold variation over the range of compounds studied.

Vicinal $^3J_{NH}$ couplings that have been studied in rigid bicyclic systems are presented in Table 27.[304] Proton couplings with ^{14}N are often difficult to

TABLE 27. $^3J_{NH}$ VALUES IN RIGID BICYCLIC RING SYSTEMS RELATED TO NORBORNANE AND BICYCLO[2.2.2]OCTANE

Compound	Nitrogen Atom Position	Proton Position	$^3J_{NH}$ (Hz)	Reference
X = OH, Y = H$_{2n}$	3n	4	ca. 0	405
	3n	2n	|1.4|	
X = H$_{2x}$, Y = OH	3n	4	ca, 0	405
	3n	2x	≤ |0.3|	
	7	8s	|2.7|	405
	7	1	≤ |0.3|	
	7	8a	|0·8|	
	5x	6x	|2.8|	299
	5x	4	|0.3|	
	5x	6n	|0.8|	
	5	8x	0	299
	5	8n	|2.5|	
	5	1	|3.5|	
	7	5x	0	299
	7	5n	|3.0|	

observe, primarily because of rapid quadrupolar relaxation of the ^{14}N nucleus, which results in broadening of the signal of the protons to which it is coupled. However, quadrupolar broadening is not a serious problem in situations where the electric field gradient at the ^{14}N nucleus is highly symmetrical (e.g., in symmetrically substituted amines, ammonium salts, and isonitriles).[299,405,442–445]

Terui et al.[405] obtained evidence for a Karplus-type relation between vicinal ^{14}N–^{1}H couplings and dihedral angle in rigid bicyclic trimethylammonium bromide salts. Based on their observations, they estimated that $^3J(^{14}$N–^{1}H) = ca. 5–6 Hz for $\phi = 180°$.[405] The results of a more recent study, which involved a variety of bicyclic ammonium salts,[299] suggest that this estimate may be too high; the value of $^3J(^{14}$N–^{1}H) = ca. 3.5 Hz was obtained in this study for $\phi = 180°$.

A number of vicinal ^{31}P–^{1}H couplings have been studied in rigid bicyclic systems. Some representative $^3J_{PH}$ values appear in Table 28. Some evidence has accrued suggesting that the quantitative angular dependence of vicinal ^{31}P–^{1}H couplings to a large extent may be a function of the state of hybridization of the phosphorus atom (i.e., whether the phosphorus atom coupled to vicinal hydrogen is tricoordinated, tetracoordinated, or pentacoordinated). White and Verkade[401] measured vicinal ^{31}P–O–C–^{1}H couplings in twenty monocyclic and polycyclic phosphites, phosphates, and thiophosphates, noting that "separate correlations" were required for tricoordinated and pentacoordinated phosphorus systems. More recently, Benezra[338] observed a Karplus-type variation of vicinal ^{31}P–C–C–^{1}H couplings with dihedral angle in a series of bicyclic and tricyclic phosphonates (pentacoordinated ^{31}P). Unfortunately, vicinal ^{31}P–C–C–^{1}H couplings for corresponding tricoordinated and tetracoordinated ^{31}P-containing compounds were not included in this study.

Interestingly, the view that vicinal ^{31}P–C–C–^{1}H coupling constants are sensitive to the phosphorus atom coordination number and to the substitution pattern around phosphorus is not universally held.[447,448] Accordingly, the issue must be regarded as being controversial at present; the situation in this regard awaits further experimental and theoretical clarification.

Carbon–carbon coupling constants in general, until relatively recently, were regarded as "spectroscopic curiosities."[280] However, vicinal carbon–carbon couplings now have been investigated fairly extensively. Representative $^3J_{CC}$ values measured in a variety of specifically ^{13}C-labeled rigid bicyclic systems are presented in Table 29. The dihedral angle dependence of $^3J_{CC}$ coupling constants appears to be considerably more complicated than is that for $^3J_{HH}$. Barfield et al. have pointed out that a Karplus-type expression "does not, in general, provide an adequate representation of the conformational dependence" of vicinal carbon–carbon coupling constants.[285] Initial experimental observations by Marshall and co-workers[286] of $^3J_{CC}$ couplings in a series of ^{13}C-carboxyl-labeled aliphatic carboxylic acids suggested the existence of a correlation with dihedral angle which was similar to that noted earlier for $^3J_{FF}$ as a function of ϕ (where $^3J_{min}$ and $^3J_{max}$ occur at values somewhat below the "Karplus values" of $\phi = 90°$ and $\phi = 180°$, respectively).[387,417]

TABLE 28. $^3J_{PH}$ Values in Rigid Bicyclic Ring Systems Related to Norbornane and Bicyclo[2.2.2]octane

Compound	Phosphorus Atom Position	Proton Position	$^3J_{PH}$ (Hz)[a]	Reference
X = H$_{5n}$	5x	4	+8.5	338
X = OH	5x	4	+7.5	338
	5x	6x	+16.5	
	5x	6n	+6.0	
X = H$_{5x}$	5n	4	0	338
	5n	6x	|8.4|	379
X = OH	5n	4	< +0·5	338
X = H$_{2n}$	2x	4	+9.	338
X = OH	2x	4	+9.	338
X = H$_{2x}$, Y = H$_{7s}$	2n	1	< +0.5	338
X = OH, Y = H$_{7s}$	2n	1	< +0.5	338
X = OH, Y = Br	2n	1	< +0.5	338

TABLE 28 (*continued*)

TABLE 28 (continued)

Compound	Phosphorus Atom Position	Proton Position	$^3J_{PH}$ (Hz)[a]	Reference
X = Y = Cl	5n	6x	+ 8.1 ± 0.14	338
	5n	6n	+ 17.9 ± 0.13	
X = Y = Cl	5n	6x	+ 8.44	380
	5n	6n	+ 17.86	
X = Y = OMe	5n	6x	|7.6|	379
	5n	6n	|17.4|	
X = H_7s, Y = Cl	5n	6x	+ 8.6 ± 0.05	338
	5n	6n	+ 17.1 ± 0.05	
	5n	6x	|7.8|	423

X = Y = Z = Ph	1	3	|7.|	446
X = Z = CMe_3, Y = Me	1	3	|7.|	446
X = Z = Me, Y = Ph	1	3	|14.|	446

X = Ph, Y = P(OMe)_2 (=O)	3a	10n	|17.4|	345
X = P(OMe)_2 (=O), Y = Me	3s	10n	|6.0|	345
X = Me, Y = P(OMe)_2 (=O)	3a	10n	|18.8|	345

TABLE 28 (continued)

TABLE 28 (*continued*)

Compound	Phosphorus Atom Position	Proton Position	$^3J_{PH}$ (Hz)[a]	Reference
X = Ph, Y = $\overset{\overset{O}{\|}}{P}(OMe)_2$	3a	10n	|19.0|	345
X = $\overset{\overset{O}{\|}}{P}(OMe)_2$, Y = Me	3s	10n	|6.0|	345
X = Me, Y = $\overset{\overset{O}{\|}}{P}(OMe)_2$	3a	10n	|18.4|	345

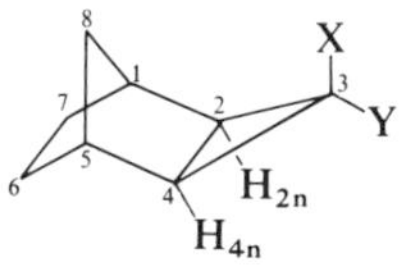

X = Ph, Y = $\overset{\overset{O}{\|}}{P}(OMe)_2$	3a	2n	|15.5|	345
X = $\overset{\overset{O}{\|}}{P}(OMe)_2$, Y = Ph	3s	2n	|4.2|	345
X = Me, Y = $\overset{\overset{O}{\|}}{P}(OMe)_2$	3a	2n	|15.0|	345
X = $\overset{\overset{O}{\|}}{P}(OMe)_2$, Y = Me	3s	2n	|3.6|	345

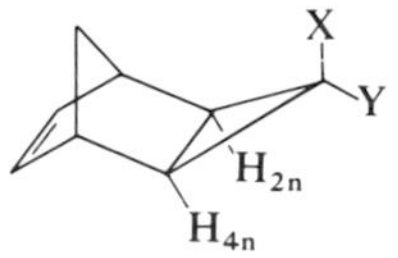

X = Ph, Y = $\overset{\overset{O}{\|}}{P}(OMe)_2$	3a	2n	|14.8|	345
X = $\overset{\overset{O}{\|}}{P}(OMe)_2$, Y = Ph	3s	2n	|4.0|	345
X = Me, Y = $\overset{\overset{O}{\|}}{P}(OMe)_2$	3a	2n	|15.0|	345
X = $\overset{\overset{O}{\|}}{P}(OMe)_2$, Y = Me	3s	2n	|3.5|	345

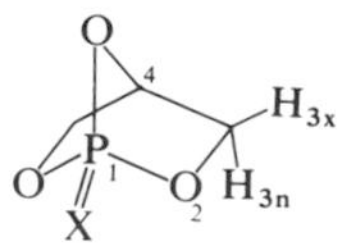

TABLE 28 (*continued*)

TABLE 28 (continued)

Compound	Phosphorus Atom Position	Proton Position	$^3J_{PH}$ (Hz)^a	Reference
X = unshared electron	1	3x	|3.83|	401
pair	1	3n	|0.29|	
	1	4	|15.9|	
X = O	1	4	|24.|	

X = Y = unshared electron pair	4	2	+2.5	401
X = S, Y = unshared electron pair	4	2	+7.5	401
X = S, Y = O	4	2	+8.2	401
X = Y = O	4	2	+8.3	401

X = unshared electron pair, R = Me	4	2	|2.|	401
X = O, R = Me	4	2	|6.|	401
X = S, R = Me	4	2	|7.|	401
X = unshared electron pair, R = H$_1$	4	2	|1.6|	401
X = O, R = H$_1$	4	2	|6.5|	401

X = H$_5$	5	4	+ 5. ± 1.	338
X = OH	5	4	+6.8	
	5	6s	+16.9	
	5	6a	+5·5	

| 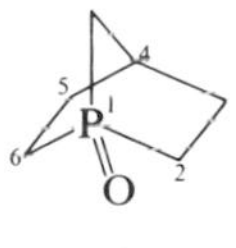 | 1 | 4 | \|28.\| | 306 |

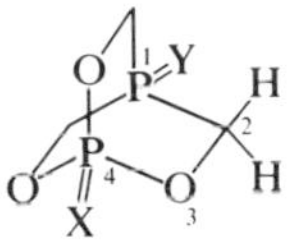

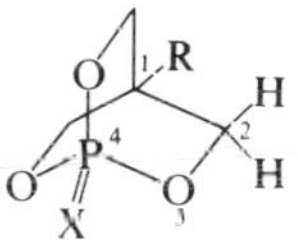

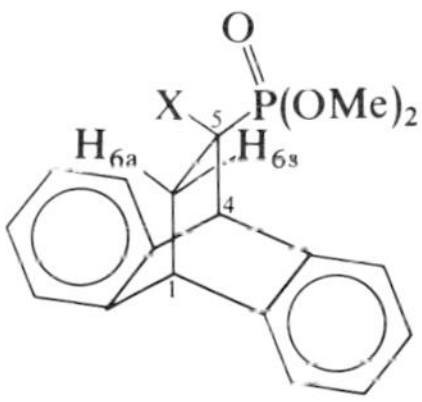

---

^{a} The absolute sign of $^3J_{PCCH}$ in compounds of the type RP(O)(OR')$_2$ has been shown to be positive.[338,382]

TABLE 29. $^3J_{CC}$ Values in Rigid Bicyclic Ring Systems Related to Norbornane and Bicyclo[2.2.2]octane

Compound[a,b]	Carbon Atom Positions	$^3J_{CC}$ (Hz)	Reference
99	2,8	$\lvert 1.04 \pm 0.1 \rvert$	287
	5,8	$\lvert 4.15 \pm 0.1 \rvert$	
100	4,8	$< \lvert 0.6 \rvert$	288, 289
	6,8	$\lvert 3.1 \rvert$	
	7,8	$\lvert 3.7 \rvert$	
	8,(CH$_3$)$_{3n}$	$\lvert 3.1 \rvert$	
	1,8^c	$\lvert 1.2 \rvert$	
	5,8^c	$< \lvert 1. \rvert$	
101	4,8	$< \lvert 0.6 \rvert$	288, 289
	6,8	$\lvert 3.1 \rvert$	
	7,8	$\lvert 4.9 \rvert$	
	8,(CH$_3$)$_{3x}$	$\lvert 2.2 \rvert$	
	1,8^c	$\lvert 0.6 \rvert$	
	5,8^c	$< \lvert 1. \rvert$	
102	4,8	$\lvert 0.7 \pm 0.2 \rvert$	288, 289
	6,8	$\lvert 0.56 \pm 0.14 \rvert$	
	7,8	$\lvert 2.7 \pm 0.1 \rvert$	
	8,(CH$_3$)$_{3n}$	$\lvert 2.8 \rvert$	
	1,8^c	$\lvert 0.5 \rvert$	
	$\lvert 5,8^c$	$< \lvert 0.2 \rvert$	
103	4,8	$\lvert 0.75 \pm 0.1 \rvert$	288, 289
	6,8	$\lvert 0.36 \pm 0.14 \rvert$	
	7,8	$\lvert 2.7 \pm 0.1 \rvert$	
	8,(CH$_3$)$_{3x}$	$\lvert 2.6 \rvert$	
	1,8^c	$\lvert 0.7 \rvert$	
	5,8^c	$< \lvert 0.2 \rvert$	
	7,8	$\lvert 4.5 \rvert$	288, 289
	8,(CH$_3$)$_{3n}$	$\lvert 4.6 \rvert$	
	7,8	$\lvert 5.2 \rvert$	288, 289
	8,(CH$_3$)$_{3n}$	$\lvert 1.9 \rvert$	
	4,9	$\lvert 2.82 \pm 0.04 \rvert$	289
	6,9	$\lvert 2.31 \pm 0.04 \rvert$	
	7,9	$\lvert 2.16 \pm 0.04 \rvert$	
	8,9	$\lvert 2.38 \pm 0.04 \rvert$	

TABLE 29 (*continued*)

TABLE 29 (continued)

Compound[a,b]	Carbon Atom Positions	$^3J_{CC}$ (Hz)	Reference
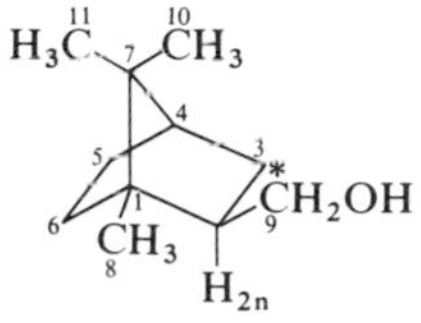 H₃C—CH₃ ... CH₂OH structure	4,9	|1.39|	289
	6,9	|3.37|	
	7,9	|0.10|	
	8,9	|2.38|	
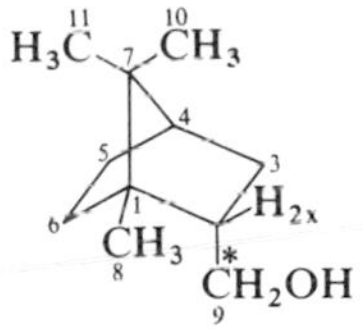 H₃C—CH₃ ... CH₂OH structure	4,9	|1.26|	289
	6,9	|3.81|	
	7,9	|2.69|	
	8,9	< |0.18|	
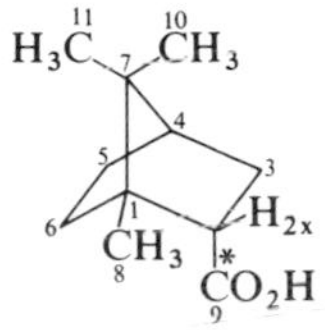 H₃C—CH₃ ... CO₂H structure	4,9	|1.45 ± 0.04|	289
	6,9	|2.22 ± 0.10|	
	7,9	|4.25 ± 0.02|	
	8,9	< |0.22|	
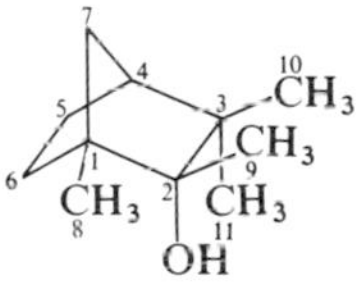 structure with OH, CH₃ positions	4,9[d]	|0.63 ± 0.05|	289
	4,9[e]	|0.73 ± 0.20|	
	4,9	|0.8|	285
	6,9[d]	|0.97 ± 0.07|	289
	6,9[e]	|0.99 ± 0.06|	
	6,9	|0.9|	285
	7,9[d]	< |0.2|	289
	7,9[e]	< |0.4|	
	7,9	|0.4|	285
	10,9[d]	|5.32 ± 0.07|	289
	10,9[e]	|5.36 ± 0.06|	
	10,9	|5.4|	285
	11,9[d]	< |0.2|	289
	11,9[e]	< |0.4|	
	11,9	< |0.4|	285
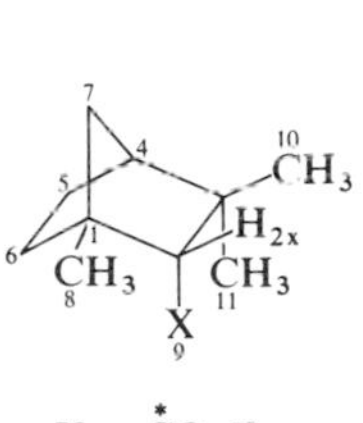 X = CO₂H	4,9	|1.95 ± 0.05|	289
	6,9	|1.06 ± 0.04|	
	7,9	|5.96 ± 0.05|	
	10,9	|0.98 ± 0.12|	
	11,9	|2.09 ± 0.04|	

TABLE 29 (continued)

TABLE 29 (*continued*)

Compound[a,b]	Carbon Atom Positions	$^3J_{CC}$ (Hz)	Reference
X = $\overset{*}{C}H_2OH$	4,9	$\lvert 1.32 \pm 0.11 \rvert$	289
	6,9	$\lvert 2.99 \pm 0.06 \rvert$	
	7,9	$\lvert 3.64 \pm 0.11 \rvert$	
	10,9	$\lvert 0.61 \pm 0.06 \rvert$	
	11,9	$\lvert 4.07 \pm 0.06 \rvert$	

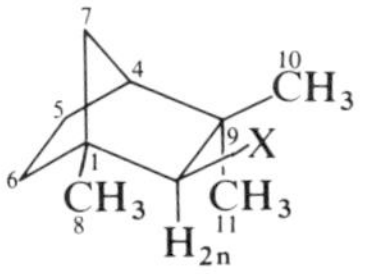

Compound[a,b]	Carbon Atom Positions	$^3J_{CC}$ (Hz)	Reference
X = $\overset{*}{C}O_2H$	4,9	$\lvert 0.78 \pm 0.05 \rvert$	289
	6,9	$\lvert 4.78 \pm 0.12 \rvert$	
	7,9	$< \lvert 0.6 \rvert$	
	10,9	$\lvert 3.35 \pm 0.04 \rvert$	
	11,9	$\lvert 1.39 \pm 0.04 \rvert$	
X = $\overset{*}{C}H_2OH$	4,9	$\lvert 0.88 \pm 0.11 \rvert$	289
	6,9	$\lvert 2.97 \pm 0.11 \rvert$	
	7,9	$< \lvert 0.3 \rvert$	
	10,9	$\lvert 5.76 \pm 0.06 \rvert$	
	11,9	$\lvert 0.68 \pm 0.06 \rvert$	
	4,8	$\lvert 3.9 \rvert$	291
	4,8	$\lvert 3.4 \rvert$	291
	1,4	$\lvert 12.2 \rvert$	291
	1,4	$\lvert 12.7 \rvert$	291

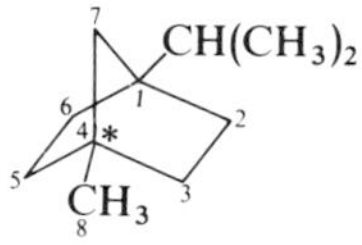

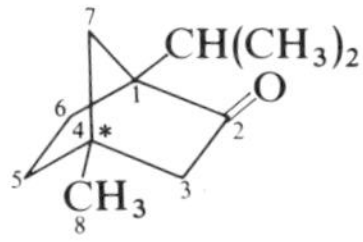

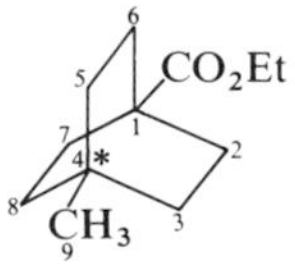

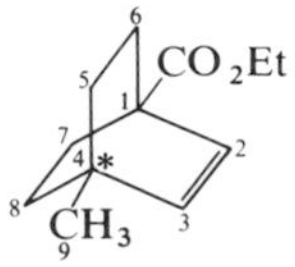

TABLE 29 (*continued*)

TABLE 29 (continued)

Compound[a,b]	Carbon Atom Positions	$^3J_{CC}$ (Hz)	Reference
X = CO₂H (see structure)			
X = $\overset{*}{C}O_2H$	3,8	|3.81 ± 0.05|	289, 383
X = $\overset{*}{C}H_2OH$	3,8	|3.51 ± 0.05|	289, 383
(structure with CH₃, OH)	4,8	|0.9|	285
	6,8	|1.8|	
	7,8	< |0.35|	
(structure with H₅ₓ, H₆ₓ, CH₃, CH₂OH)	2,8	< |0.5 ± 0.1|	287
	4,8	< |0.5 ± 0.1|	
	7,8	|4.5 ± 0.1|	
	8,(CH₃)₅ₙ	|4.57 ± 0.1|	
X = $\overset{*}{C}O_2H$	3,9	|3.52 ± 0.05|	289, 383
X = $\overset{*}{C}H_2OH$	3,9	|3.39 ± 0.08|	289, 383
104	2,5	|1.4|	292
	2,8	|2.4|	
105	2,5	|1.8|	292
	2,8	|2.7|	
106	2,5	|2.2|	292
	2,8	|2.2|	
107	2,5	|2.4|	292
	2,8	|2.4|	
108	2,5	|1.7|	292
	2,8	|1.8|	

[a] See Table 10 for structures **96–108**.
[b] Asterisk indicates position of specific ^{13}C labeling.
[c] $^3J_{CC}$ coupling can occur via a pathway involving the ether oxygen atom in compounds **100–103**.[288,289]
[d] CDCl₃ solvent.
[e] (CD₃)₂C=O solvent.

Subsequent experimental and theoretical (INDO-FPT) studies of $^3J_{CC}$ coupling constants[285,409,410] revealed that the magnitude (absolute value) of $^3J_{CC}$ couplings in substituted butanes generally exceed corresponding $^3J_{CC}$ couplings in rigid bicyclic systems. To account for this, Barfield and co-workers[409] suggested the operation of an additional spin–spin coupling mechanism in the rigid bicyclic systems studied (e.g., substituted adamantanes and norbornanes) that is absent in the (acyclic) substituted butanes. This additional mechanism, termed "impinging multiple rear-lobe effects,"[409] was pictured in terms of a nonbonded (direct) orbital rear-lobe overlap effect. This effect was thought to be "analogous to the mechanism which leads to long range proton–proton coupling over four bonds in the 'W' arrangement of a propanic fragment, but of opposite sign."[409] However, a more recent INDO-MO theoretical study of substituent effects upon $^3J_{CC}$[449,450] questioned the necessity of invoking these orbital rear-lobe overlap effects. Instead, the discrepancies noted above were explained simply in terms of β-substituent effects. Subsequently, Barfield and co-workers[383,411,438,439] utilizing a modified INDO-FPT procedure, identified "γ-substituent effects arising from 1,4-type interactions"[438] as being capable of providing a substantial negative contribution to overall $^3J_{CC}$ values in substituted norbornanes and adamantanes.

In an effort to further clarify the conformational dependence of $^3J_{CC}$ coupling constants in alicyclic compounds, Berger[291] studied a series of similarly substituted rigid bicyclic systems bearing site-specific skeletal ^{13}C atom enrichment. These systems possess the advantage of conformational rigidity but minimize the possibility of through-space interactions between the two coupling carbons (believed to be a source of difficulty when interpreting $^3J_{CC}$ couplings in, e.g., ^{13}C-carboxyl-labeled ortho-substituted benzoic acids[451]). In the absence of such complicating factors, Berger[291] found that the angular dependence of $^3J_{CC}$ couplings in skeletally ^{13}C-labeled bicyclic systems could be expressed quantitatively by a Karplus-type relation: $^3J_{CC} = 1.67 + 0.176 \cos \phi + 2.24 \cos 2\phi$. This relation was obtained "assuming additivity of ^{13}C–^{13}C coupling constants transmitted by different pathways."[333] (The assumption of approximate algebraic additivity of J_{CC} couplings appears to be reasonably well grounded).[452]

Some representative vicinal carbon–fluorine coupling constants that have been measured in specifically fluorinated norbornanes and bicyclo[2.2.2]-octanes are presented in Table 30. Although a relatively small number of such systems has been studied, there is some evidence that the magnitude of $^3J_{FC}$ may be a function of dihedral angle.[80] In acyclic systems [e.g., in $FCH_2CH_2CH_3$ and in $FCH_2CH(OH)CH_3$], a clear conformational dependence of $^3J_{FC}$ is indicated on the basis of the results of INDO-MO calculations.[453] These calculations predict $^3J_{FC}$ to be of opposite sign to $^1J_{FC}$ (i.e., positive), to be dominated by the Fermi contact term, and to show "an angular dependence similar to that found for $^3J_{HH}$ and $^3J_{FH}$."[453]

Table 31 presents $^3J_{NC}$ coupling constants that have been measured in bicyclic ammonium salts.[293,294] Berger[296] has stated that vicinal ^{15}N–^{13}C

TABLE 30. $^3J_{FC}$ Values in Rigid Bicyclic Ring Systems Related to Norbornane and Bicyclo[2.2.2]octane

Compound	Fluorine Atom Position	Carbon Atom Position	$^3J_{FC}$ (Hz)[a]	Reference
(norbornane structure, F_{2x}, H_{2n})	2x	4	|2.3|	80
	2x	6	|9.8|	
	2x	7	<|1.|	
anti–*syn*, *exo*, *endo* (F_{2x}, F_{2n}) structure				
R = H	2x,2n	4	|5.4, 2.8|	80
	2x,2n	6	|5.8, 5.8|	
	2x,2n	7	|4.9|	
R = 1-CH_3	2x,2n	4	|4.5, 2.6|	80
	2x,2n	6	|6.3, 4.2|	
	2x,2n	7	|5.3|	
	2x,2n	1-CH_3	|4.0|	
R = *exo*-3-CH_3	2x,2n	4	|5.6, 1.2|	80
	2x,2n	6	|5.8, 5.8|	
	2x,2n	7	|4.7|	
	2x,2n	$(CH_3)_{3x}$	|14.1, 2.7|	
R = *endo*-3-CH_3	2x,2n	4	|4.4, 1.9|	80
	2x,2n	6	|6.9, 5.9|	
	2x,2n	7	|5.8|	
	2x,2n	$(CH_3)_{3n}$	|1.5, 9.7|	
R = *exo*-5-CH_3	2x,2n	4	|3.2, 3.2|	80
	2x,2n	6	|6.0, 6.0|	
	2x,2n	7	|5.0|	
R = *endo*-5-CH_3	2x,2n	4	|4.2, 2.2|	80
	2x,2n	6	|6.0, 6.0|	
	2x,2n	7	|4.0|	
R = *exo*-6-CH_3	2x,2n	4	|4.6, 2.0|	80
	2x,2n	6	|5.8, 5.8|	
	2x, 2n	7	|5.0|	
R = *endo*-6-CH_3	2x,2n	6	|3.4, 3.4|	80
	2x,2n	7	|5.2|	
R = *syn*-7-CH_3	2x,2n	4	|4.0, 3.0|	80
	2x,2n	6	|8.1, 6.8|	
	2x,2n	7	|5.5|	
R = *anti*-7-CH_3	2x,2n	4	|3.5, 2.6|	80
	2x,2n	6	|5.8, 5.8|	
	2x,2n	7	|4.3|	
(bicyclo[2.2.2]octane structure, F)	1	3	|8.2|	277
	1	4	|7.9|	

[a] $^3J_{FC}$ values are considered to be accurate to ± 0.5 Hz.[80]

TABLE 31. $^3J_{NC}$ Values in Rigid Bicyclic Ring Systems Related to Norbornane and Bicyclo[2.2.2]octane

Compound[a]	Nitrogen Atom Position	Carbon Atom Position	$^3J_{NC}$ (Hz)[b]	Reference
109	2n	4	$< |0.5|^c$	293
	2n	$(CH_3)_1$	$< |0.2|^c$	
	2n	$(CH_3)_{3x}$	$< |0.2|^c$	
	2n	$(CH_3)_{3n}$	$< |0.2|^c$	
	2n	7	$|1.8|^c$	
110	1	4	$|3.5|^c$	293
111	1	4	$|4.8|^c$	293
112	1	4	$|2.8|$	294
113	1	4	$|6.7|$	294

[a] See Table 11 for structures 109–113.
[b] Unless otherwise indicated, couplings are presented for ^{13}C directly bonded to ^{14}N. An asterisk indicates specific ^{15}N labeling and, in these instances, the couplings given are for ^{13}C directly bonded to ^{15}N.
[c] Estimated from line widths.

coupling constants are "not well-suited as a conformational probe (vis-à-vis $^3J_{CC}$) since vicinal $^{15}N–^{13}C$ spin coupling constants are difficult to resolve." However, Anteunis and co-workers[293] were able to observe vicinal $^{14}N–^{13}C$ couplings in broad-band proton-decoupled ^{13}C NMR spectra of tetraalkylammonium ions 109–111 (as well as in other tetraalkylammonium ions). These investigators[293] found a Karplus-type dependence of vicinal $^{14}N–^{13}C$ coupling constants on dihedral angle. The orientation dependence of $^3J_{NC}$ in these ammonium salts could be described satisfactorily by the Karplus curve calculated by Solkan and Bystrov[415] for $^3J_{NC}$ values in peptides. However, it was recognized that "other factors" (in addition to dihedral angle) contributed to the overall magnitude of $^3J_{NC}$.[293] Vicinal $^{15}N–^{13}C$ coupling constants also have been studied in ^{15}N-labeled quinuclidine and in its hydrochloride salt.[294]

Vicinal $^{31}P–^{13}C$ coupling constants measured in rigid bicyclic phosphines,[111] phosphonates[310,311] phosphine oxides,[308] phosphine sulfides,[111,307] and phosphonium salts[111] are presented in Table 32. Some evidence has accrued suggesting that the question of the possible conformational dependence of $^3J_{PC}$ coupling constants must be approached with consideration given to the oxidation (hybridization) state of the ^{31}P nucleus. Quin and Littlefield[111] established the existence of a Karplus-type dihedral angle dependence for $^3J_{PC}$ in syn- and anti-7-norbornenyl phosphine sulfides and phosphonium salts. However, $^3J_{PC}$ coupling constants measured in corresponding norbornenyl systems bearing tricoordinated phosphorus functions [e.g., Cl_2P and $(CH_3)_2P$] in the 7 position failed to display a Karplus-type dependence upon dihedral angle.[111] [It may be recalled that a similar situation was encountered by White and Verkade[401] in their study of the conformational dependence of vicinal $^{31}P–O–C–^1H$ couplings (see above, in this section)]. However, the results of a recent investigation of ^{31}P and ^{13}C NMR spectra of

TABLE 32. $^3 J_{PC}$ Values in Rigid Bicyclic Ring Systems Related to Norbornane and Bicyclo[2.2.2]octane

Compound	Phosphorus Atom Position	Carbon Atom Position	$^3 J_{PC}$ (Hz)	Reference
X = Cl$_2$P	7	2,3	|10.4|	111
	7	5,6	|6.1|	
X = (CH$_3$)$_2$P	7	2,3	|10.|	111
	7	5,6	|7.|	
	2n	4	|3.9|	310
	2n	6	|7.3|	
	2n	7	|15.2|	
	2n	(CH$_3$)$_{10}$	|0.0|	
	2n	S(CH$_3$)$_{11}$	|0.0|	
	2x	4	|1.8|	311
	2x	6	|18.4|	
	2x	7	< |0.6|	
X = Cl$_2$P	7	2,3	|6.7|	111
	7	5,6	< |1.|	
X = (CH$_3$)$_2$P	7	2,3	|3.7|	111
	7	5,6	|2.6|	
X = Cl$_2$P	7	2,3	|6.2|	111
	7	5,6	|9.7|	
X = (CH$_3$)$_2$P	7	2,3	|6.1|	111
	7	5,6	|8.6|	
X = (CH$_3$)$_2$P ∥ S	7	2,3	|15.9|	111
	7	5,6	|0.|	

TABLE 32 (continued)

TABLE 32 (*continued*)

Compound	Phosphorus Atom Position	Carbon Atom Position	$^3J_{PC}$ (Hz)	Reference
X = (CH$_3$)$_3$P$^+$ I$^-$	7	2,3	|16.5|	111
	7	5,6	|0.|	
(norbornyl phosphorus structure, P=O)	1	4	|47.|	308
(structure: S, H$_{7s}$, PPh$_2$, Ph$_2$P, H$_{8a}$, H$_3$C, H$_{5a}$, CH(CH$_3$)$_2$)	7a	2	|14.|	307
	8s	5	|14.|	
(structure: P(OCH$_3$)$_2$, Ph)	3s	1,5	|1.1|	311
(structure: Ph, P(OCH$_3$)$_2$, O)	3a	1,5	|6.1|	311

2-norbornyl phosphorus compounds suggest that it may be possible to accommodate $^3J_{PC}$ couplings in terms of a Karplus-type dihedral angle dependence, especially were the signs of several data points in a plot of $^3J_{PC}$ vs. ϕ to be "reinterpreted."[414] On the basis of their findings, the authors (Quin et al.) concluded that their new information "definitely established that, for P(IV) functions in general, a rather ordinary plot of $^3J_{PC}$ vs. ϕ is obtained, but that for P(III) functions, the $^3J_{PC}$ maxima at 0° and 180° are greatly different (about 35 and 10 Hz, respectively). Furthermore, the minimum, with an inversion of coupling sign, occurs about 15° higher [for P(III) functions] than the usual value of 90°."[414] Thiem and Meyer[310] have verified independently the existence of a Karplus-type relationship for vicinal ^{31}P–C–C–^{13}C couplings in phosphonates.

Vicinal fluorine–fluorine couplings constants are presented in Table 33.[387] It appears that the magnitudes of $^3J_{FF}$ couplings are influenced strongly by conformational factors. However, their relationship is complex and cannot be described simply in terms of a Karplus-type relation.[387,418,455] A theoretical

TABLE 33. $^3J_{FF}$ Values in Rigid Bicyclic Ring Systems Related to Norbornane and Bicyclo[2.2.2]octane[a,b]

Compound	Fluorine Atom Positions	$^3J_{FF}$ (Hz)	Reference
X = OCH$_3$, Y = F, Z = Br	4,5x	|8.4|	378
	4,5n	|3.0|	
	4,7s	|8.4|	
	4,7a	|3.3|	
	5x,6x	|1.8|	
	5x,6n	|2.7|	
	5n,6x	|1.5|	
	5n,6n	|2.8|	
X = F, Y = CH$_3$, Z = I	2,3	|5.0|	378
	5x,6x	|1.4|	
	5x,6n	|2.9|	
	5n,6x	|1.4|	
	5n,6n	|2.1|	
X = OCH$_3$, Y = F, Z = I	4,5x	|8.8|	378
	4,5n	|3.9|	
	4,7s	|8.6|	
	4,7a	|3.8|	
	5x,6x	|2.2|	
	5x,6n	|1.0|	
	5n,6x	|1.3|	
	5n,6n	|1.5|	
X = F, Y = Z = CH$_3$	5x,6x	|1.5|	378
	5x,6n	|2.2|	
	5n,6x	|1.8|	
	5n,6n	|2.4|	
	5x,6x	|12.5|	363
	5x,6x	ca. |9.5|	363

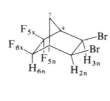

TABLE 33 (*continued*)

TABLE 33 (*continued*)

Compound	Fluorine Atom Positions	$^3J_{FF}$ (Hz)	Reference

Compound	Fluorine Atom Positions	$^3J_{FF}$ (Hz)	Reference
X = H	7,8s	−1.2	418
	7,8a	+9.4	
X = Cl	7,8s	−2.2	418
	7,8a	+11.4	

[a] An extensive tabulation of $^3J_{FF}$ values in rigid bicyclic systems is presented in ref. 275.
[b] The absolute sign of $^3J_{FF}$ can be either positive or negative in fluorinated alkenes, and it is probably negative in fluorinated alkanes.[388,464]

study of the angular dependence of vicinal $^{19}F–C–C–^{19}F$ couplings has revealed that " the total vicinal coupling constant is mainly affected by the Fermi contact term, changing its sign twice between 0° and 180°."[387] In another study, Williamson and Braman found "no obvious correlation between $^3J_{FF}$ and dihedral angle"[454] in a series of 3-halo-1,1-dichloro-2,2,3-trifluorocyclopropanes.

Comparison of data in Tables 33 and 44 reveals that $^3J_{FF}$ couplings in highly fluorinated norbornenes are, in general, considerably smaller than are the corresponding $^4J_{FF}$ couplings in these systems. In an effort to account for the seemingly anomalous behavior of vicinal (and long-range) fluorine–fluorine couplings in these and other fluorinated compounds, Sederholm and co-workers[456–458] suggested that the mechanism of transmission of spin–spin coupling information between two fluorine nuclei may involve both through-space and through-bond contributions. The relative importance of through-space and through-bond mechanisms in this regard has been the subject of considerable controversy. The through-space mechanism in spin–spin coupling has been examined in detail in a recent review by Hilton and Sutcliffe[459] (cf. discussion in the next section). Despite the considerable amount of attention that $^3J_{FF}$ couplings have received, the need for further experimental and theoretical clarification of the factors (e.g., conformational factors and substituent effects) which determine the sign and magnitude of $^3J_{FF}$ coupling constants is apparent.*

* A referee commented: "You have hit upon a very tender nerve in the discussion of 'through-bond' and 'through-space' coupling ... My suggestion ... would be to emphasize the confusion associated with ill-defined terminology. Clearly, you are correct in saying that there is need for further experimental and theoretical work on F–F coupling."

The remaining vicinal coupling constants that have been studied in rigid bicyclic systems, i.e., $^3J_{PF}$,[460] $^3J_{PP}$,[448,461] and miscellaneous $^3J_{XY}$,[29,207–213,290] are presented in Tables 34–36. Vicinal $^{31}P-^{31}P$ couplings have been studied in endo Diels-Alder dimers of phospholium salts (134),[461] phosphole oxides (135),[461] and phosphole sulfides (136).[448] Vicinal $^{31}P-^{31}P$ couplings in com-

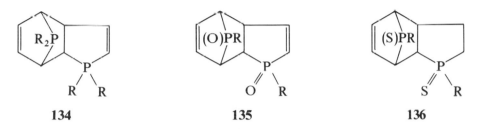

pounds related to 134 and 135 are relatively large (35–45 Hz).[461] Interestingly, $^3J_{PP}$ in exo Diels-Alder dimer 137a was found to be essentially zero, whereas $^3J_{PP}$ in the corresponding endo dimer (137b) was observed to be 45 Hz. The large $^3J_{PP}$ coupling observed in 137b was thought to have "a conformational origin."[448]

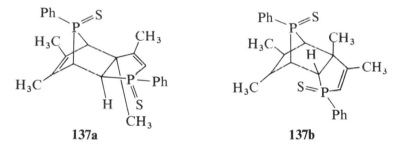

Karplus-type dependence of $^3J_{MC}$ (M = metal atom) upon dihedral angle has been observed in a number of rigid bicyclic organometallic compounds (e.g., norbornyltin,[316] mercury,[317] and thallium[319] compounds).

TABLE 34. $^3J_{PF}$ VALUES IN RIGID BICYCLIC RING SYSTEMS RELATED TO NORBORNANE AND BICYCLO[2.2.2]OCTANE

Compound	Phosphorus Atom Position	Fluorine Atom Position	$^3J_{PF}$ (Hz)	Reference
	1	2-(CF$_3$)	|40.|	460

Compound	Phosphorus Atom Positions	$^3J_{PP}$ (Hz)	Reference
Diels-Alder dimer of			

R = H	Bridging P, 2-phospholene P	|38.|	461
R = OCH$_3$	Bridging P, 2-phospholene P	|44.|	461
Diels-Alder dimer of			

R = H	Bridging P, 2-phospholene P	|35.|	461
R = OCH$_3$	Bridging P, 2-phospholene P	|36.|	461
Diels-Alder dimer of			

R = PhCH$_2$, X = Br	Bridging P, 2-phospholene P	|36.|	461
R = CH$_3$, X = I	Bridging P, 2-phospholene P	|44.|	461
Diels-Alder dimer of			

	Bridging P, 2-phospholene P	|40.|	461

	3,10	ca. 0	448

	3,10	|45.|	448

TABLE 36. MISCELLANEOUS $^3J_{XY}$ VALUES IN RIGID BICYCLIC RING SYSTEMS RELATED TO NORBORNANE AND BICYCLO[2.2.2]OCTANE

Compound	X Atom Position	Y Atom Position	$^3J_{XY}$ (Hz)	Reference
(structure: norbornane, Sn(CH$_3$)$_3$, H$_{2n}$)	$^{119}Sn_{1x}$	$^{13}C_4$	|34.9|	316
	$^{119}Sn_{2x}$	$^{13}C_6$	|34.0|	
	$^{119}Sn_{2x}$	$^{13}C_7$	|38.6|	
(structure: norbornane, H$_{2x}$, Sn(CH$_3$)$_3$)	$^{119}Sn_{2n}$	$^{13}C_4$	|36.9|	316
	$^{119}Sn_{2n}$	$^{13}C_6$	|30.1|	
	$^{119}Sn_{2n}$	$^{13}C_7$	|40.9|	
(structure: norbornane, HgOAc, H$_{2n}$)	$^{199}Hg_{2x}$	$^{13}C_4$	|93.|	317
	$^{199}Hg_{2x}$	$^{13}C_6$	|276.|	
	$^{199}Hg_{2x}$	$^{13}C_7$	|9.|	
(structure: norbornane, H$_{2x}$, HgOAc)	$^{199}Hg_{2n}$	$^{13}C_4$	|93.|	317
	$^{199}Hg_{2n}$	$^{13}C_6$	|159.|	
	$^{199}Hg_{2n}$	$^{13}C_7$	|244.|	
(structure: OAc, HgCl, H$_{3n}$, H$_{2n}$)	$^{199}Hg_{2x}$	$^{13}C_4$	|10.|	317
	$^{199}Hg_{2x}$	$^{13}C_6$	|256.|	
	$^{199}Hg_{2x}$	$^{13}C_7$	|12.|	
(structure: OAc, Tl(OAc)$_2$, H$_{3n}$, H$_{2n}$)	$^{205}Tl_{2x}$	$^{13}C_1$	|27.|	319[a]
	$^{205}Tl_{2x}$	$^{13}C_6$	|1301.|	
	$^{205}Tl_{2x}$	$^{13}C_7$	|32.|	
(structure: OAc, Tl(OAc)$_2$, H$_{6n}$, H$_{5n}$)	$^{205}Tl_{5x}$	$^{13}C_1$	<|3.|	319[a]
	$^{205}Tl_{5x}$	$^{13}C_3$	|1057.|	
	$^{205}Tl_{5x}$	$^{13}C_7$	|17.|	
(structure: CH$_3$, CH$_3$, CH$_2$, HgCl)	$^{199}Hg_1$	$^{13}C_3$	|102.5|	322[b]

TABLE 36 (*continued*)

TABLE 36 (*continued*)

Compound	X Atom Position	Y Atom Position	$^3J_{XY}$ (Hz)	Reference
(structure: H₃C, H₃C, CH₂, Hg; positions 1–7)	$^{199}Hg_1$	$^{13}C_3$	$\lvert 80.9 \rvert$	322[b]
(structure: ClHg, OAc, H₃ₙ, H₅ₙ)	$^{199}Hg_{3x}$	$^{13}C_1$	$\lvert 35. \rvert$	317
	$^{199}Hg_{3x}$	$^{13}C_5$	$\lvert 314. \rvert$	
	$^{199}Hg_{3x}$	$^{13}C_6$	$\lvert 14. \rvert$	
	$^{199}Hg_{3x}$	$^{13}C_7$	$\lvert 41. \rvert$	
(structure: OAc, Tl(OAc)₂, H₃ₙ, H₂ₙ)	$^{205}Tl_{2x}$	$^1H_1$	$\lvert 649. \rvert$	590[c]
	$^{205}Tl_{2x}$	$^1H_{3n}$	$\lvert 624. \rvert$	

(structure: (AcO)₂Tl, H₆ₓ, H₅ₙ, R, R′, H₂ₓ, O, O; positions 1,4,7)

Compound	X Atom Position	Y Atom Position	$^3J_{XY}$ (Hz)	Reference
$R = H_{3x}$, $R' = H_{3n}$	$^{203/205}Tl_{5x}$	$^1H_4$	$\lvert 630. \rvert$	389
	$^{203/205}Tl_{5x}$	$^1H_{6x}$	$\lvert 909. \rvert$	
$R = H_{3x}$, $R' = CO_2H$	$^{203/205}Tl_{5x}$	$^1H_4$	$\lvert 515. \rvert$	389
	$^{203/205}Tl_{5x}$	$^1H_{6x}$	$\lvert 911. \rvert$	
$R = CO_2H$, $R' = H_{3n}$	$^{203/205}Tl_{5x}$	$^1H_4$	$\lvert 545. \rvert$	389
	$^{203/205}Tl_{5x}$	$^1H_{6x}$	$\lvert 898. \rvert$	

(structure: R, R′, OAc, Z, H₆ₙ, H₅ₙ; positions 1,2,3,4,7)

Compound	X Atom Position	Y Atom Position	$^3J_{XY}$ (Hz)	Reference
$R = R' = H$, $Z = Tl(OAc)_2$	$^{205}Tl_{5x}$	$^{13}C_1$	$\lvert 45. \rvert^a$	318
			$\lvert 59. \rvert^b$	319
	$^{205}Tl_{5x}$	$^{13}C_3$	$\lvert 1149. \rvert^a$	318
			$\lvert 1101. \rvert^b$	319
	$^{205}Tl_{5x}$	$^{13}C_7$	$\lvert 7. \rvert^a$	318
			$\lvert 5. \rvert^b$	319

TABLE 36 (*continued*)

TABLE 36 (*continued*)

Compound	X Atom Position	Y Atom Position	$^3J_{XY}$ (Hz)	Reference
R = H, R′ = OCH$_3$, Z = Tl(OAc)$_2$	^{205}Tl$_{5x}$	^{13}C$_1$	|54.|	318
	^{205}Tl$_{5x}$	^{13}C$_3$	|1154.|	
	^{205}Tl$_{5x}$	^{13}C$_7$	<|3.|	
R = H, R′ = Cl, Z = Tl(OAc)$_2$	^{205}Tl$_{5x}$	^{13}C$_1$	|46.|	318
	^{205}Tl$_{5x}$	^{13}C$_7$	|10.|	
R = Cl, R′ = H, Z = Tl(OAc)$_2$	^{205}Tl$_{5x}$	^{13}C$_1$	|49.|	318
	^{205}Tl$_{5x}$	^{13}C$_3$	|1157.|	
	^{205}Tl$_{5x}$	^{13}C$_7$	|10.|	

Compound	X Atom Position	Y Atom Position	$^3J_{XY}$ (Hz)	Reference
L = H$_{3x}$, M = H$_{3n}$, N = H$_{2x}$, R = Tl(OAc)$_2$	^{205}Tl$_{5x}$	^{13}C$_1$	|78.|	318
	$^{203/205}$Tl$_{5x}$	^{13}C$_1$	|75.7|	320
	^{205}Tl$_{5x}$	^{13}C$_3$	|1282.|	318
	$^{203/205}$Tl$_{5x}$	^{13}C$_3$	|1289.|	320
	^{205}Tl$_{5x}$	^{13}C$_7$	|127.|	318
	$^{203/205}$Tl$_{5x}$	^{13}C$_7$	|131.8|	320
L = H$_{3x}$, M = CO$_2$CH$_3$, N = H$_{2x}$, R = Tl(OAc)$_2$	^{205}Tl$_{5x}$	^{13}C$_1$	|108.|	318
	^{205}Tl$_{5x}$	^{13}C$_3$	|1111.|	
	^{205}Tl$_{5x}$	^{13}C$_7$	|149.|	
L = H$_{3x}$, M = CO$_2$CH$_3$, N = CH$_3$, R = Tl(OAc)$_2$	^{205}Tl$_{5x}$	^{13}C$_1$	|71.|	318
	^{205}Tl$_{5x}$	^{13}C$_3$	|1111.|	
	^{205}Tl$_{5x}$	^{13}C$_7$	|169.|	
L = H$_{3x}$, M = H$_{3n}$, N = H$_{2x}$, R = HgOAc	^{199}Hg$_{5x}$	^{13}C$_1$	|17.1|	320
	^{199}Hg$_{5x}$	^{13}C$_3$	|437.|	
	^{199}Hg$_{5x}$	^{13}C$_7$	|36.6|	
L = H$_{3x}$, M = H$_{3n}$, N = H$_{2x}$, R = HgCl	^{199}Hg$_{5x}$	^{13}C$_3$	|440.|	320

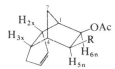

Compound	X Atom Position	Y Atom Position	$^3J_{XY}$ (Hz)	Reference
R = HgOAc	^{199}Hg$_{5x}$	^{13}C$_1$	|15.|	321
	^{199}Hg$_{5x}$	^{13}C$_3$	|212.|	
R = Tl(OAc)$_2$	^{205}Tl$_{5x}$	^{13}C$_1$	|83.|	321
	^{205}Tl$_{5x}$	^{13}C$_3$	|1124.|	
	^{205}Tl$_{5x}$	^{13}C$_7$	|12.|	

TABLE 36 (*continued*)

TABLE 36 (*continued*)

Compound	X Atom Position	Y Atom Position	$^3J_{XY}$ (Hz)	Reference
	$^{205}Tl_{5x}$	$^{13}C_7$	|12.|	321
	$^{199}Hg_{5x}$	$^{13}C_3$	|291.|	321
	^{119}Sn	$^{13}C_5$	|50.8|	462

a CDCl$_3$ solvent.
b Pyridine solvent.
c The sign of $^3J_{XY}$(X = ^{205}Tl, Y = ^{1}H) is opposite of that of $^2J_{XY}$ (X = ^{205}Tl, Y = ^{1}H).[50]

Four-Bond and More Distant (Long-Range) Couplings

Like the corresponding vicinal couplings discussed in the preceding section, four-bond (long-range) proton–proton couplings have been extensively utilized in conformational and configurational analysis. Experimental values for $^4J_{HH}$ couplings measured in saturated, cyclic systems are presented in a number of reviews,[463–465] and detailed theoretical studies of $^4J_{HH}$ couplings have appeared.[357,466–468] Representative $^4J_{HH}$ couplings measured in rigid bicyclic systems are presented in Table 37.[469,470]

Meinwald and Lewis[484] first utilized a rigid bicyclic ring system, *exo*-5-chlorobicyclo 2.1.1 hexane-*exo*-6-*t*-butylcarboxamide (**138**), to demonstrate the

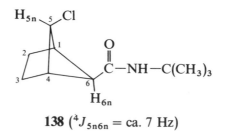

138 ($^4J_{5n6n}$ = ca. 7 Hz)

TABLE 37. LONG RANGE $^nJ_{HH}$ VALUES ($n \geq 4$) IN RIGID BICYCLIC RING SYSTEMS RELATED TO NORBORNANE AND BICYCLO[2.2.2]OCTANE[a]

Compound	Proton Positions	Value of n	$^nJ_{HH}$ (Hz)	Reference
	1,3 = 2,4	4	|1.2|	264
	1,4	4	0.	125
	2,7a = 3,7a	4	|0.5–0.6|	63, 125
	2,7s = 3,7s	4	|0.20–0.35|	125
	5n,7s = 6n,7s	4	|2.0–2.4|	63, 125
W = H_{5x}, X = H_{6x}, Y = Z = Cl	1,3 = 2,4	4	|0.65|	258
	1,4	4	0.	
	2,7a	4	< |0.5|	
W = Z = Cl, X = H_{6x}, Y = H_{5x}	1,3 = 2,4	4	|0.70|	258
	1,4	4	0.	
	2,7a	4	< |0.5|	
	5n,7s	4	|2.0|	
W = Y = Cl, X = H_{6x}, Z = H_{6n}	1,4	4	0.	258
	2,7a	4	< |0.5|	
	5n,7s	4	|3.0|	
W = H_{5x}, X = H_{6x}, Y = Cl, Z = H_{6n}	1,4	4	0.	258
	2,7a	4	< |0.5|	
	5n,7s	4	|3.1|	
W = H_{5x}, X = H_{6x}, Y = OH, Z = H_{6n}	1,4	4	0.	331
	2,7a	4	< |0.3|	
	5n,7s	4	|2.8|	
W = OH, X = H_{6x}, Y = H_{5n}, Z = H_{6n}	1,4	4	0.	331
	2,7a	4	< |0.3|	
	5n,7s	4	|2.9|	
W = H_{5x}, X = H_{6x}, Y = CO_2CH_3, Z = H_{6n}	1,4	4	0.	331
	2,7a	4	|0.3|	
	5n,7s	4	|3.0|	
W = CO_2CH_3, X = H_{6x}, Y = H_{5n}, Z = H_{6n}	1,4	4	0.	331
	2,7a	4	< |0.3|	
	5n,7s	4	|3.0|	
W = H_{5x}, X = H_{6x}, Y = CN, Z = H_{6n}	1,4	4	0.	331
	2,7a	4	|0.4|	
	5n,7s	4	|3.1|	
W = CN, X = H_{6x}, Y = H_{5n}, Z = H_{6n}	1,4	4	0.	331
	2,7a	4	|0.85|	
	5n,7s	4	|3.0|	
W = PhS=O, X = H_{6x}, Y = H_{5n}, Z = H_{6n}	5n,7s	4	|2.|	422
	6n,7s	4	|2.|	
W = $PhCH_2S$, X = H_{6x}, Y = H_{5n}, Z = H_{6n}	5n,7s	4	|2.|	422
W = PhS, X = H_{6x}, Y = H_{5n}, Z = H_{6n}	5n,7s	4	|2.|	422
W = H_{5x}, X = H_{6x}, Y = Z = Br	1,4	4	ca. |1.0|	329
	2,7a	4	ca. |0.8|	
W = X = Br, Y = H_{6n}, Z = H_{6n}	5n,7s	4	|2.1|	329

TABLE 37 (continued)

TABLE 37 (*continued*)

Compound	Proton Positions	Value of n	$^{n}J_{HH}$ (Hz)	Reference
$W = Z = Br, X = H_{6x}$, $\quad Y = H_{6n}$	4,6x 5n,7s	4 4	ca. 0. $\lvert 2.5 \rvert$	329
$W = Br, X = H_{6x}$, $\quad Y = H_{5n}, Z = Cl$	4,6x 5n,7s	4 4	ca. 0. $\lvert 2.3 \rvert$	329
$W = H_{5x}, X = Cl$, $\quad Y = Br, Z = H_{6n}$	4,6x 5n,7s	4 4	ca. 0. $\lvert 2.4 \rvert$	329

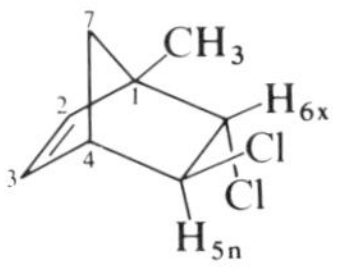	5n,7s	4	$\lvert 2.4 \rvert$	329

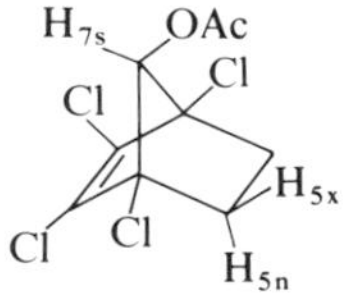	5n,7s	4	$\lvert 1.3 \rvert$	329

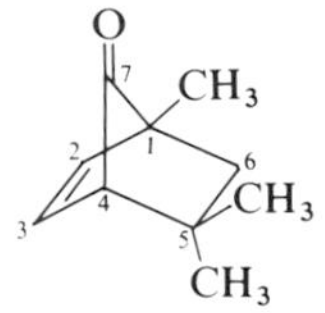	2,4	4	$\lvert 1.4 \rvert$	464

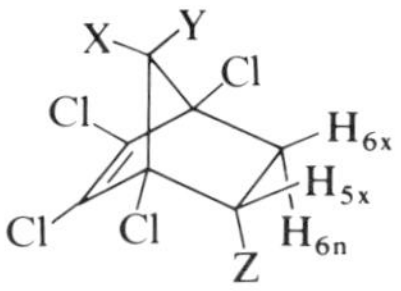 $X = H_{7s}, Y = Cl$, $\quad Z = CN$	5x,7s 6x,7s 6n,7s	4 4 4	$+0.42$ $+0.21$ $+1.49$	354
$X = H_{7s}, Y = Cl$, $\quad Z = CO_2CH_3$	5x,7s 6x,7s 6n,7s	4 4 4	$+0.45$ $+0.30$ $+1.42$	354
$X = H_{7s}, Y = Cl$, $\quad Z = OAc$	5x,7s 6x,7s 6n,7s	4 4 4	$+0.42$ $+0.23$ $+2.22$	354
$X = H_{7s}, Y = Cl$, $\quad Z = Br$	5x,7s 6x,7s 6n,7s	4 4 4	$+0.42$ $+0.38$ $+2.08$	354
$X = H_{7s}, Y = Z = Cl$	5x,7s 6x,7s 6n,7s	4 4 4	$+0.48$ $+0.38$ $+2.15$	354
$X = H_{7s}, Y = Cl$, $\quad Z = P(O)(OCH_3)_2$	6n,7s	4	$\lvert 1.4 \pm 0.04 \rvert$	338

TABLE 37 (*continued*)

TABLE 37 (*continued*)

Compound	Proton Positions	Value of n	$^nJ_{HH}$ (Hz)	Reference
$X = H_{7s}$, $Y = Cl$, $Z = Ph$	5x,7s	4	+0.40	354
	6x,7s	4	+0.22	
	6n,7s	4	+1.68	
$X = H_{7s}$, $Y = Cl$, $Z = OAc$	5x,7s	4	|0.4|	355
	6x,7s	4	|0.2|	
	6n,7s	4	|2.2|	
$X = H_{7s}$, $Y = Z = OAc$	6n,7s	4	|2.1|	355
$X = H_{7s}$, $Y = H_{7a}$, $Z = OAc$	6n,7s	4	|3.5|	355
	6n,7a	4	|0.9|	
	5x,$(CH_3)_{6n}$	4	0.32	357
	5n,$(CH_3)_{6n}$	4	+0.15	357
	5n,$(CH_3)_{6x}$	4	−0.10	
	6x,$(CH_3)_{5n}$	4	−0.13	357
	6n,$(CH_3)_{5n}$	4	−0.17	
$W = H_{5x}$, $X = H_{6x}$, $Y = (CH_3)_{5n}$, $Z = (CH_3)_{6n}$	5x,$(CH_3)_{6n}$	4	−0.32	357
$W = H_{5x}$, $X = (CH_3)_{6x}$, $Y = (CH_3)_{5n}$, $Z = H_{6n}$	5x,$(CH_3)_{6x}$	4	+0.15 or −0.10	357
	6n,$(CH_3)_{5n}$	4	−0.10 or +0.15	
$W = H_{5x}$, $X = H_{6x}$, $Y = H_{5n}$, $Z = (CH_3)_{6n}$	5x,$(CH_3)_{6n}$	4	−0.13	357
	5n,$(CH_3)_{6n}$	4	−0.17	

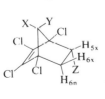

TABLE 37 (*continued*)

TABLE 37 (*continued*)

Compound	Proton Positions	Value of n	$^nJ_{HH}$ (Hz)	Reference
$X = H_{7s}$, $Y = Z = Cl$	5x,7s	4	$\|0.37\|$	473
	6x,7s	4	$\|0.32\|$	
	6n,7s	4	$\|2.12\|$	
$X = Z = Cl$, $Y = H_{7a}$	5x,7a	4	ca. 0.	473
	6x,7a	4	ca. 0.	
	6n,7a	4	ca. 0.	
$X = H_{7s}$, $Y = Cl$, $Z = F$	5x,7s	4	$\|0.30\|$	473
	6x,7s	4	ca. 0.	
	6n,7s	4	$\|2.32\|$	
$X = Cl$, $Y = H_{7a}$, $Z = F$	5x,7a	4	ca. 0.	473
	6x,7a	4	ca. 0.	
	6n,7a	4	ca. 0.	

$X = OCH_3$	6x,$(CO_2CH_3)_{5n}$	6	$\|0.18\|$	474
$X = Cl$	6x,$(CO_2CH_3)_{5n}$	6	$\|0.18\|$	474
$X = Br$	6x,$(CO_2CH_3)_{5n}$	6	$\|0.17\|$	474
$X = I$	6x,$(CO_2CH_3)_{5n}$	6	$\|0.16\|$	474
$X = CH_3$	6x,$(CO_2CH_3)_{5n}$	6	$\|0.21\|$	474
$X = H_{6n}$	6x,$(CO_2CH_3)_{5n}$	6	$\|0.27\|$	474

$X = OCH_3$	6n,$(CO_2CH_3)_{5n}$	6	$\|0.29\|$	474
$X = Cl$	6n,$(CO_2CH_3)_{5n}$	6	$\|0.28\|$	474
$X = Br$	6n,$(CO_2CH_3)_{5n}$	6	$\|0.28\|$	474
$X = I$	6n,$(CO_2CH_3)_{5n}$	6	$\|0.28\|$	474
$X = CH_3$	6n,$(CO_2CH_3)_{5n}$	6	$\|0.27\|$	474

$X = OCH_3$	6x,$(CO_2CH_3)_{5x}$	6	$\|0.19\|$	474
$X = Cl$	6x,$(CO_2CH_3)_{5x}$	6	$\|0.18\|$	474
$X = Br$	6x,$(CO_2CH_3)_{5x}$	6	$\|0.19\|$	474

TABLE 37 (*continued*)

TABLE 37 (continued)

Compound	Proton Positions	Value of n	$^nJ_{HH}$ (Hz)	Reference
X = I	6x,(CO$_2$CH$_3$)$_{5x}$	6	|0.19|	474
X = CH$_3$	6x,(CO$_2$CH$_3$)$_{5x}$	6	|0.18|	474
X = H$_{6n}$	6x,(CO$_2$CH$_3$)$_{5x}$	6	|0.18|	474

(structure: bicyclic alkene with X, CO$_2$CH$_3$, H$_{6n}$, H$_{5n}$)

X = OCH$_3$	6n,(CO$_2$CH$_3$)$_{5x}$	6	|0.17|	474
X = Cl	6n,(CO$_2$CH$_3$)$_{5x}$	6	|0.16|	474
X = Br	6n,(CO$_2$CH$_3$)$_{5x}$	6	|0.16|	474
X = I	6n,(CO$_2$CH$_3$)$_{5x}$	6	|0.14|	474
X = CH$_3$	6n,(CO$_2$CH$_3$)$_{5x}$	6	|0.15|	474

(structure: bicyclic with H$_{3x}$, H$_{2x}$, X, CO$_2$CH$_3$)

X = Cl	3x,(CO$_2$CH$_3$)$_{2n}$	6	|0.15|	474
X = Br	3x,(CO$_2$CH$_3$)$_{2n}$	6	|0.17|	474
X = I	3x,(CO$_2$CH$_3$)$_{2n}$	6	< |0.1|	474
X = CH$_3$	3x,(CO$_2$CH$_3$)$_{2n}$	6	|0.18|	474
X = H$_{3n}$	3x,(CO$_2$CH$_3$)$_{2n}$	6	|0.24|	474

(structure: bicyclic with X, H$_{2x}$, H$_{3n}$, CO$_2$CH$_3$)

X = OCH$_3$	3n,(CO$_2$CH$_3$)$_{2n}$	6	|0.23|	474
X = Cl	3n,(CO$_2$CH$_3$)$_{2n}$	6	|0.23|	474
X = Br	3n,(CO$_2$CH$_3$)$_{2n}$	6	|0.24|	474
X = I	3n,(CO$_2$CH$_3$)$_{2n}$	6	|0.24|	474
X = CH$_3$	3n,(CO$_2$CH$_3$)$_{2n}$	6	|0.24|	474

(structure: bicyclic with H$_{3x}$, CO$_2$CH$_3$, X, H$_{2n}$)

| X = OCH$_3$ | 3x,(CO$_2$CH$_3$)$_{2x}$ | 6 | \|0.19\| | 474 |
| X = Cl | 3x,(CO$_2$CH$_3$)$_{2x}$ | 6 | \|0.18\| | 474 |

TABLE 37 (continued)

TABLE 37 (*continued*)

Compound	Proton Positions	Value of n	$^nJ_{HH}$ (Hz)	Reference
$X = Br$	$3x,(CO_2CH_3)_{2x}$	6	$\lvert 0.18 \rvert$	474
$X = I$	$3x,(CO_2CH_3)_{2x}$	6	$\lvert 0.18 \rvert$	474
$X = CH_3$	$3x,(CO_2CH_3)_{2x}$	6	$\lvert 0.20 \rvert$	474
$X = H_{3n}$	$3x,(CO_2CH_3)_{2x}$	6	$\lvert 0.19 \rvert$	474

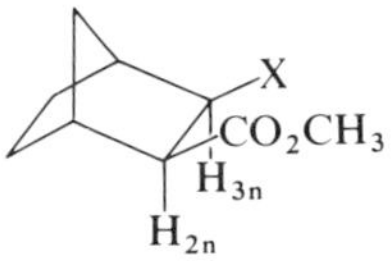				
$X = Cl$	$3n,(CO_2CH_3)_{2x}$	6	$\lvert 0.17 \rvert$	474
$X = Br$	$3n,(CO_2CH_3)_{2x}$	6	$\lvert 0.14 \rvert$	474
$X = I$	$3n,(CO_2CH_3)_{2x}$	6	$\lvert 0.13 \rvert$	474
$X = CH_3$	$3n,(CO_2CH_3)_{2x}$	6	$\lvert 0.14 \rvert$	474

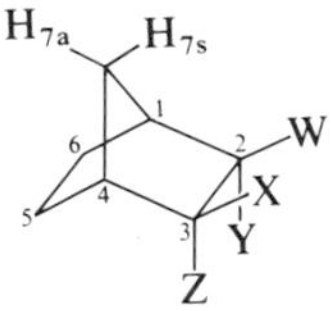

$W = X = Br, Y = H_{2n}$, $Z = H_{3n}$	$2n,7a = 3n,7a$	4	$\lvert 2.9 \rvert$	329
$W = Br, X = Cl$, $Y = H_{2n}, Z = H_{3n}$	$2n,7a = 3n,7a$	4	$\lvert 2.1 \rvert$	329
$W = Z = Br, X = H_{3x}$, $Y = H_{2n}$	$2n,7a = 3n,7a$	4	ca. $\lvert 2.9 \rvert$	329
$W = H_{2x}, X = Cl$, $Y = Br, Z = H_{3n}$	$2n,7a = 3n,7a$	4	$\lvert 2.8 \rvert$	329
$W = Br, X = H_{3x}$, $Y = H_{2n}, Z = Cl$	$2n,7a = 3n,7a$	4	$\lvert 2.6 \rvert$	329

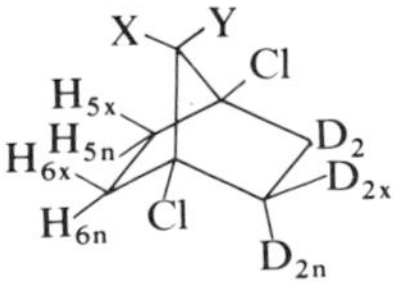

$X = Y = Cl$	$2x,6x$	4	b	341
$X = OAc, Y = H_{7s}$	$5n,7s$	4	$\lvert 1.16 \pm 0.01 \rvert$	341
$X = H_{7a}, Y = Cl$	$5n,7s$	4	-2.19 ± 0.08	341

TABLE 37 (*continued*)

TABLE 37 (*continued*)

Compound	Proton Positions	Value of n	$^nJ_{HH}$ (Hz)	Reference
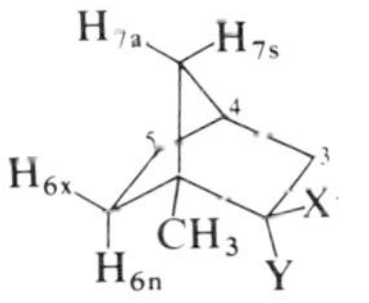				
X = H$_{2x}$, Y = OH	2x,6x	4	|1.4|	119
X = OH, Y = H$_{2n}$	2n,7a	4	|1.4|	
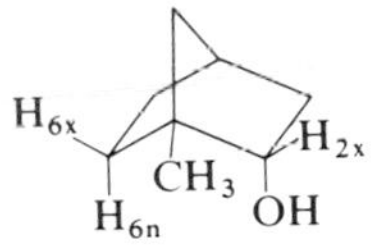	2x,6x	4	|1.8|	119
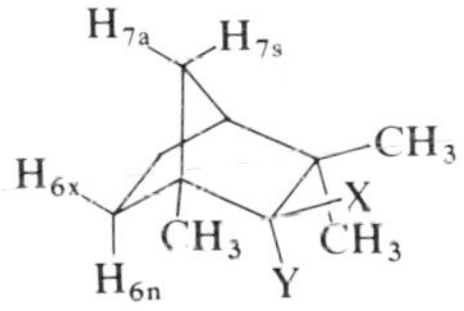				
X = H$_{2x}$, Y = OH	2x,6x	4	|1.35|	119
X = OH, Y = H$_{2n}$	2n,7a	4	|1.85|	119
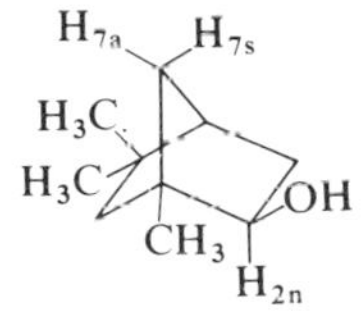	2n,7a	4	< |1.6|	119
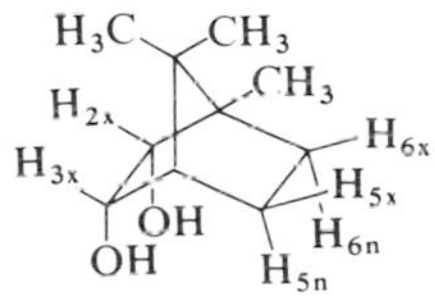	2x,6x	4	|1.4|	426
	3x,5x	4	|1.0|	
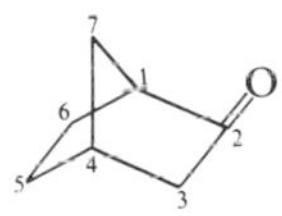	1,3n	4	0.0	342^c
	1,3x	4	0.0	
	1,4	4	+1.17	
	1,5n	4	−0.30	

TABLE 37 (*continued*)

TABLE 37 (*continued*)

Compound	Proton Positions	Value of n	$^nJ_{HH}$ (Hz)	Reference
	1,5x	4	+0.15	
	3n,5n	4	0.0	
	3n,5x	4	0.0	
	3n,6n	5	0.0	
	3n,6x	5	0.0	
	3x,5n	4	0.0	
	3x,5x	4	+2.26	
	3x,6n	5	0.0	
	3x,6x	5	0.0	
	3x,7s	4	0.0	
	3x,7a	4	0.0	
	4,6n	4	−0.45	
	4,6x	4	+0.65	
	5n,7a	4	−0.10	
	5n,7s	4	+2.09	
	5x,7a	4	0.0	
	5x,7s	4	0.0	
	6n,7a	4	−0.13	
	6n,7s	4	+2.03	
	6x,7a	4	0.0	
	6x,7s	4	0.0	

(X = halogen)

Compound	Proton Positions	Value of n	$^nJ_{HH}$ (Hz)	Reference
	3n,7a	4	ca. |3.–4.|	473

| Z = Br | 3x,5x | 4 | \|1.0\| | 476 |
| Z = OAc | 3x,5x | 4 | \|1.5\| | |

| | 4,6x | 4 | +1.1 | 359 |

TABLE 37 (*continued*)

TABLE 37 (*continued*)

Compound	Proton Positions	Value of n	$^{n}J_{HH}$ (Hz)	Reference
H_3C CH_3 / CH_3 / H_{6x} / H_{6n} / O	$6n,(CH_3)_1$	4	$\mid 0.1\text{–}0.2 \mid$	357
H_{7a} H_{7s} / F_{5x} Br / X Br / F_{5n} H_{3n} / Y H_{2n}				
X = Y = F	$2n,7a = 3n,7a$	4	$\mid 1.8 \mid$	363
X = Y = CF_3	$2n,7a \simeq 3n,7a$	4	$\mid 1.7 \mid$	
X = Y = Cl	$2n,7a \simeq 3n,7a$	4	$\mid 1.8 \mid$	
X = CF_2Cl, Y = Cl	$2n,7a \simeq 3n,7a$	4	$\mid 1.9 \mid$	
H_{7a} H_{7s} / F_2 Cl / X / F_2 H_{3n} / Y				
X = Cl, Y = H	$2n,7a \simeq 3n,7a$	4	$\mid 1.9 \mid$	365
X = Y = Cl	$3n,7a$	4	$\mid 2. \mid$	
H_{7s} H_{7a} / X / H_{2x} H_{5x} / H_{3n} Y / O H_{5n} / O				
X = I, Y = H_{6x}	$2x,4$	4	$\mid 1.0 \mid$	356
	$1,5x$	4	$\mid 0.5 \mid$	
	$3n,7a$	4	$\mid 2.6 \mid$	
	$5n,7s$	4	$\mid 1.3 \mid$	
	$2x,6x$	4	$\mid 1.2 \mid$	
	$4,6x$	4	$\mid 0.6 \mid$	
	$1,4$	4	$\mid 1.4 \mid$	
X = Br, Y = H_{6x}	$2x,4$	4	$\mid 1.0 \mid$	356
	$1,5x$	4	$\mid 0.3 \mid$	
	$3n,7a$	4	$\mid 2.4 \mid$	
	$5n,7s$	4	$\mid 2.0 \mid$	
	$2x,6x$	4	$\mid 1.0 \mid$	
	$1,4$	4	$\mid 1.6 \mid$	

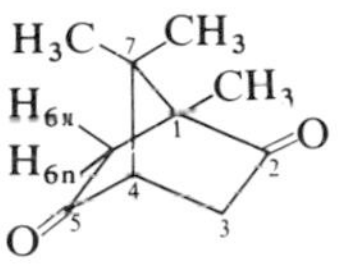

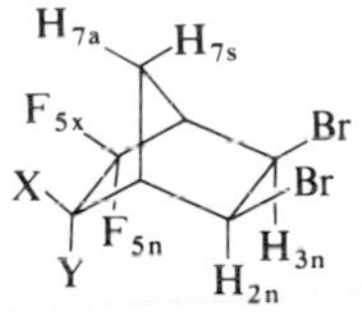

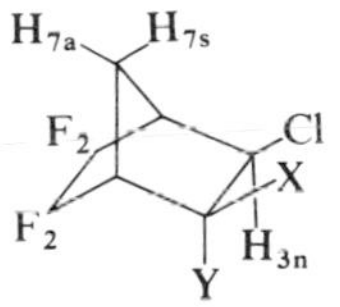

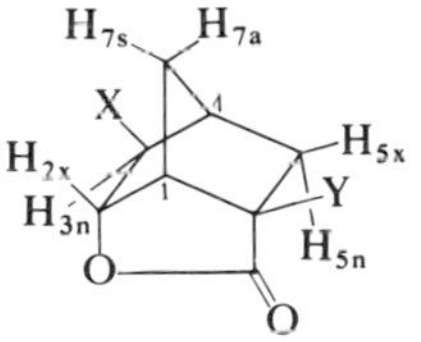

TABLE 37 (*continued*)

TABLE 37 (*continued*)

Compound	Proton Positions	Value of n	$^nJ_{HH}$ (Hz)	Reference
X = OAc, Y = H$_{6x}$	2x,4	4	|1.2|	356
	1,5x	4	|0.3|	
	3n,7a	4	|1.7|	
	2x,6x	4	|1.2|	
	1,4	4	|1.2|	
X = OTs, Y = H$_{6x}$	2x,4	4	|1.1|	356
	3n,7a	4	|1.7|	
	2x,6x	4	|1.1|	
	1,4	4	|1.0|	
X = I, Y = (CH$_3$)$_{6x}$	1,5x	4	|0.8|	356
	3n,7a	4	|2.1|	
	5n,7s	4	|2.1|	
	1,4	4	|1.4|	
X = Br, Y = (CH$_3$)$_{6x}$	2x,4	4	|1.0|	356
	3n,7a	4	|2.1|	
	5n,7s	4	|1.9|	
	1,4	4	|1.4|	

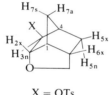

X = OTs	5n,7s	4	|2.2|	356
X = OAc	3n,7a	4	|1.8|	356
	5n,7s	4	|2.2|	
	2x,6x	4	|1.0|	

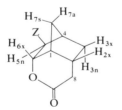

X = I	2x,6x	4	|1.8|	361
	3n,7s	4	|2.5|	
	5n,7a	4	|3.2|	
X = Br	2x,6x	4	|2.0|	361
	3n,7s	4	|2.2|	
	5n,7a	4	|3.0|	
X = HgOAc	2x,6x	4	|2.2|	361
	3n,7s	4	|2.5|	
	5n,7a	4	|3.7|	

TABLE 37 (*continued*)

TABLE 37 (*continued*)

Compound	Proton Positions	Value of n	$^nJ_{HH}$ (Hz)	Reference
X = HgBr	2x,6x	4	|2.0|	361
	3n,7s	4	|2.5|	
	5n,7a	4	|3.5|	

| 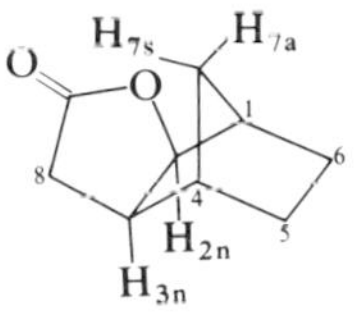 | 2n,4 | 4 | ca. \|1.\| | 361 |
| | 2n,7a | 4 | ca. \|1.\| | |

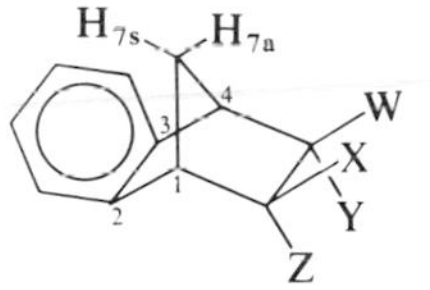 W = S—⟨ ⟩—CH₃,	5n,7s	4	|2.5|	360
X = H_6x, Y = H_5n, Z = Cl				
W = SiCl₃, X = D,	5n,7s	4	|1.5|	360
Y = H_5n, Z = H_6n	6n,7s	4	|1.5|	
W = X = D, Y = H_5n,	5n,7s	4	|2.0|	360
Z = H_6n	6n,7s	4	|2.0|	
W = CCl₃, X = H_6x,	5n,7s	4	|1.5|	360
Y = H_5n, Z = Br				
W = CCl₃, X = H_6x,	5n,7s	4	|1.6|	360
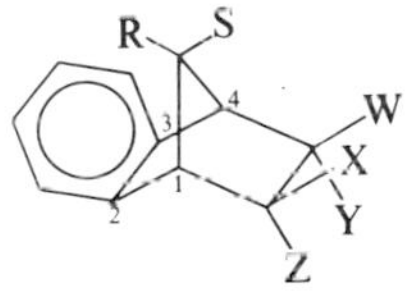 Y = H_5n, Z = Cl				

R = H_7s, S = W = Br,	5n,7s	4	|1.2–1.4|	362
X = H_6x, Y = H_5n,	6n,7s	4	|1.2–1.4|	
Z = H_6n				
R = H_7s, S = H_7a,	6n,7s	4	|2.4|	362
W = H_5x, X = Y = Cl,				
Z = H_6n				
R = H_7s, S = Y = Cl,	6n,7s	4	|2.0|	362
W = H_5x, X = H_6x,				
Z = H_6n				

TABLE 37 (*continued*)

TABLE 37 (*continued*)

Compound	Proton Positions	Value of n	$^nJ_{HH}$ (Hz)	Reference
$R = H_{7s}$, $S = H_{7a}$, $W = X = Cl$, $Y = H_{5n}$, $Z = H_{6n}$	5n,7s 6n,7s	4 4	|1.9| |1.9|	362
$R = H_{7s}$, $S = W = Cl$, $X = H_{6x}$, $Y = H_{5n}$, $Z = H_{6n}$	5n,7s 6n,7s	4 4	|1.2–1.3| |1.2–1.3|	362
$R = H_{7s}$, $S = Cl$, $W = OAc$, $X = H_{6x}$, $Y = H_{5n}$, $Z = H_{6n}$	5n,7s 6n,7s	4 4	|1.3–1.4| |1.3–1.4|	362

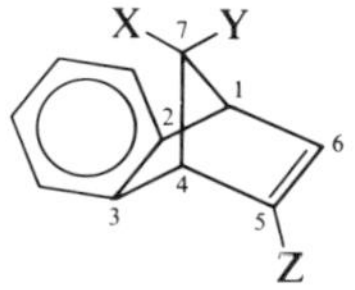 Compound	Proton Positions	Value of n	$^nJ_{HH}$ (Hz)	Reference
$X = H_{7s}$, $Y = H_{7a}$, $Z = H_5$	1,5	4	|1.2|	264
$X = H_{7s}$, $Y = OH$, Cl, or Br, $Z = H_5$	5,7s = 6,7s	4	|0.8–0.9|	362
$X = H_{7s}$, $Y = H_{7a}$, $Z = H_5$ or Cl	5,7s $\simeq$ 6,7s	4	|0.8–0.9|	362
$X = Cl$, $Y = H_{7a}$, $Z = H_5$	5,7s = 6,7s	4	|0.8–0.9|	362

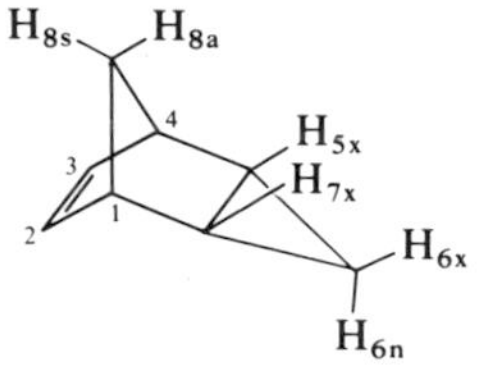	6x,8s	5	|2.3|	477

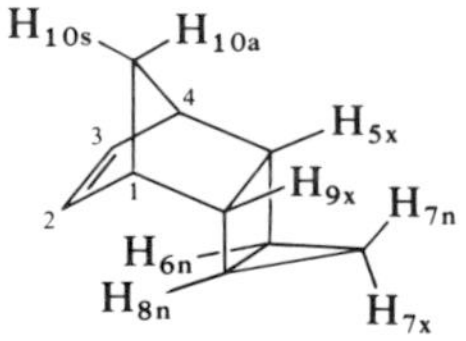	7x,10s	6	|1.|	478

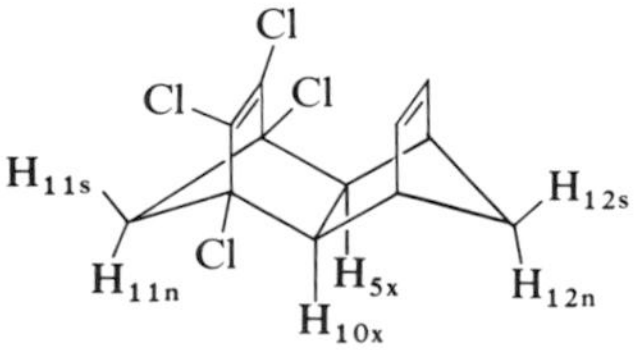	11s,12s	6	|1.|	479

TABLE 37 (*continued*)

TABLE 37 (*continued*)

Compound	Proton Positions	Value of n	$^{n}J_{HH}$ (Hz)	Reference
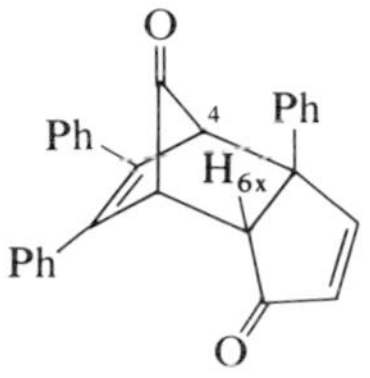	1,4	4	$|1.5|$	427
	4,6x	4	0.	
$X = H_7$	1,3	4	$|0.95 \pm 0.1|$	431
	1,3	4	$|1.1|$	264
	2,7	4	0.	431
$X = Cl$	2,7	4	$|0.8|$	220^c
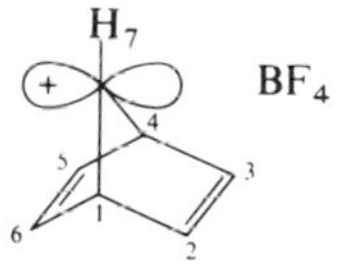	1,3	4	$|1.5|$	220^c
	1,4	4	$|0.5|$	
	1,5	4	$|1.8|$	
	2,7	4	$|2.7|$	
	6,7	4	$|1.0|$	
$X = H_2, H_3$	1,3	4	$|1.3|$	124
	1,8	4	$|1.3|$	
$X = CO_2CH_3$	1,5	4	$+1.6$	
$W = H_{5n}, X = H_{6n},$	1,3	4	$|1.5|$	124
$Y = H_{5x}, Z = H_{6x}$	1,4	5	$< |0.4|$	
$W = X = Y = Z = CN$	1,3	4	$+1.3$	

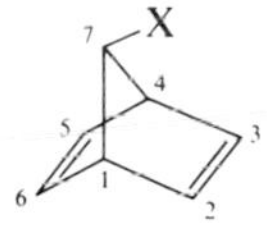

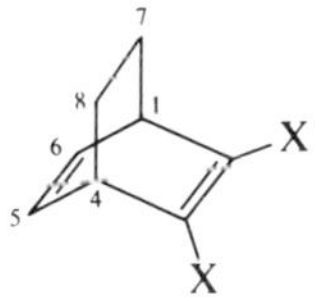

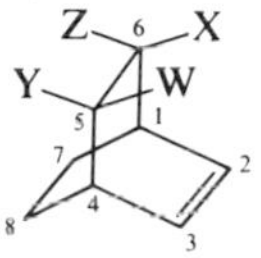

TABLE 37 (*continued*)

TABLE 37 (*continued*)

Compound	Proton Positions	Value of n	$^nJ_{HH}$ (Hz)	Reference
W, X 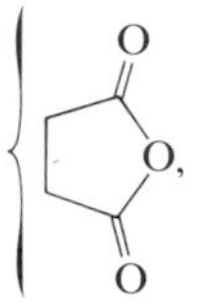 =	1,3	4	|1.7|	124
Y = H$_{5x}$, Z = H$_{6x}$ W = X = CO$_2$CH$_3$, Y = H$_{5x}$, Z = H$_{6x}$	1,3	4	|1.7|	124
W = H$_{5n}$, X = Y = CO$_2$CH$_3$, Z = H$_{6x}$	1,3	4	|1.8|	124
W = H$_{5n}$, X = OH, Y = H$_{5x}$, Z = H$_{6x}$	1,3	4	|1.6|	124
W = H$_{5n}$, X = OAc, Y = H$_{5x}$, Z = H$_{6x}$	1,3	4	|1.6|	124

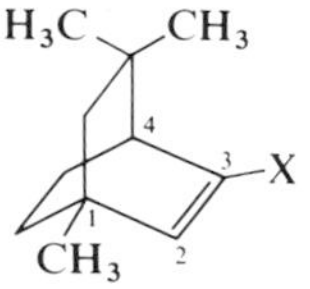				
X = H$_3$	2,4	4	+ 1.30	480[d]
X = N(CH$_3$)$_2$	2,4	4	|1.90|	480
X = OAc	2,4	4	|1.85|	480
X = Cl	2,4	4	|1.75|	480
X = I	2,4	4	|1.63|	480
X = Ph	2,4	4	ca. |1.4|	480
X = H$_3$C—C=O	2,4	4	|1.40|	480
X = CO$_2$H	2,4	4	|1.30|	480
X = CO$_2$CH$_3$	2,4	4	|1.30|	480

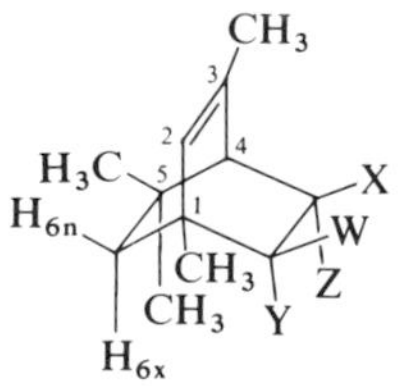				
W, X = , Y = H$_{7x}$,	2,4	4	|1.5|	135
	2,(CH$_3$)$_3$	4	|1.8|	

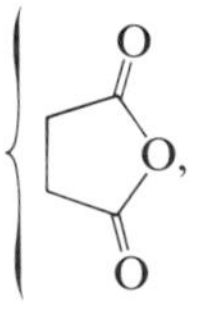

TABLE 37 (*continued*)

TABLE 37 (*continued*)

Compound	Proton Positions	Value of n	$^nJ_{HH}$ (Hz)	Reference
$Z = H_{8x}$				
$W = X = CO_2CH_3$, $Y = H_{/x}$,	2,4	4	$|1.7|$	135
$Z = H_{8x}$	$2,(CH_3)_3$	4	$|1.6|$	
$W = H_{7n}$, $X = Y = CO_2CH_3$,	2,4	4	$|1.7|$	135
$Z = H_{8x}$	$2,(CH_3)_3$	4	$|1.6|$	
	6n,7n	4	$|2.0|$	
$W = Z = CO_2CH_3$, $X = H_{8n}$,	2,4	4	$|1.5|$	135
$Y = H_{7x}$	$2,(CH_3)_3$	4	$|1.8|$	

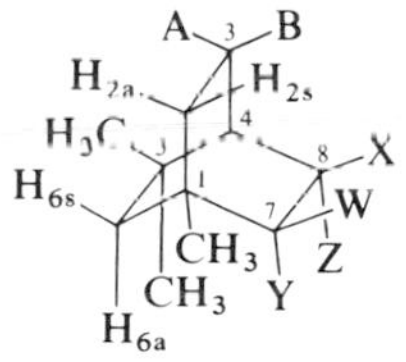

$A = H_{3a}$, $B = (CH_3)_{3s}$,	2a,7a	4	$|2.3|$	135
	3a,8a	4	$|1.2|$	

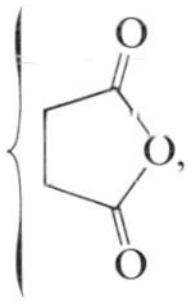

$Y = H_{7a}$, $Z = H_{8a}$				
$A = H_{3a}$, $B = (CH_3)_{3s}$,	2a,7a	4	$|2.2|$	135
$W = X = CO_2CH_3$,	3a,8a	4	$|1.8|$	
$Y = H_{7a}$, $Z = H_{8a}$				
$A = (CH_3)_{3a}$, $B = H_{3s}$,	6s,7s	4	$|1.6|$	135
$W = H_{7s}$, $X = Y = CO_2CH_3$,				
$Z = H_{8a}$				
$A = H_{3a}$, $B = (CH_3)_{3s}$,	2a,7a	4	$|1.8|$	135
$W = Z = CO_2CH_3$,				
$X = H_{8s}$, $Y = H_{7n}$				

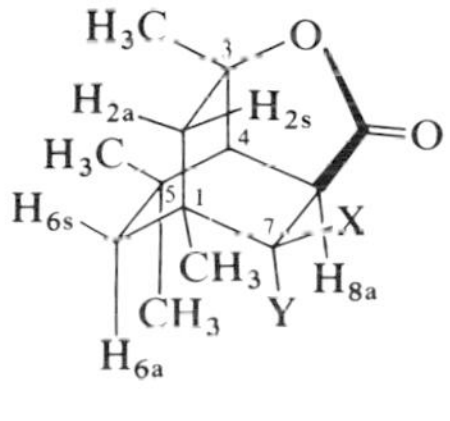

$X = CO_2CH_3$, $Y = H_{7a}$	2a,7a	4	$|1.5|$	135
	2s,6a	4	$|1.6|$	
$X = H_{7s}$, $Y = CO_2CH_3$	6s,7s	4	$|2.0|$	135

TABLE 37 (*continued*)

TABLE 37 (*continued*)

Compound	Proton Positions	Value of n	$^nJ_{HH}$ (Hz)	Reference
$W = X = CO_2CH_3$, $Y = H_{5x}, Z = H_{6x}$	1,5x	4	$\lvert 0.8 \rvert$	135
$W = H_{5n}, X = H_{6n}$, $Y = H_{5x}, Z = OH$	6n,7n	4	$\lvert 1.2 \rvert$	135
$W = H_{5n}, X = H_{6n}$, $Y = H_{5x}, Z = OAc$	6n,7n	4	$\lvert 1.2 \rvert$	135
$W = H_{7s}, X = H_{8s}$, $Y = H_{7a}, Z = H_{8a}$	1,5	4	$\lvert 1.6 \rvert$	124
$W = H_{7s}, X = H_{8s}$, $Y = Z = CO_2CH_3$	1,5	4	$\lvert 1.3 \rvert$	124
	1,8s	4	$\lvert 0.9 \rvert$	
$W = Z = CO_2CH_3$, $X = H_{8s}, Y = H_{7a}$	1,4	5	$\lvert 1.8 \rvert$	124
$W = X = CO_2CH_3$, $Y = H_{7a}, Z = H_{8a}$	1,8a	4	$\lvert 1.0 \rvert$	124
$W = Y = H_7, X = Z = H_8$	1,8	4	$\lvert 1.3 \rvert$	124
$W = X = CH_3, Y = H_7, Z = H_8$	7,(CH$_3$)$_8$	4	-0.36	357
$W = H_7, X = Y = CH_3, Z = H_8$	7,(CH$_3$)$_8$	4	-0.02	357
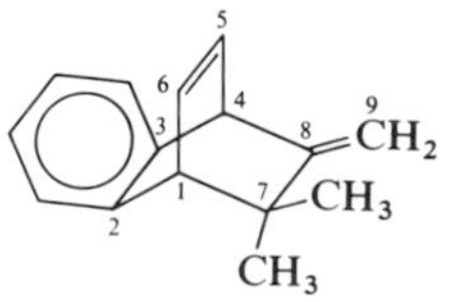	1,5	4	$+1.62$	357
	4,6	4	$+1.21$	

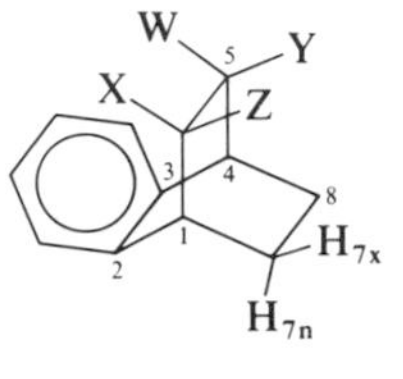

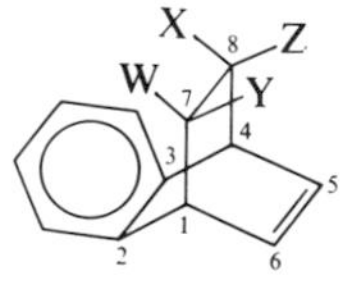

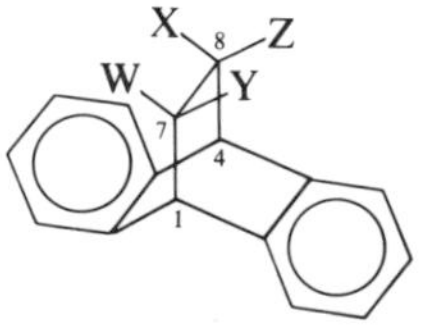

TABLE 37 (*continued*)

TABLE 37 (*continued*)

Compound	Proton Positions	Value of n	$^nJ_{HH}$ (Hz)	Reference
	5,7s	4	$+0.64 \pm 0.03$	483
	5,7a	4	-0.15 ± 0.03	
	3x,5x	4	$\lvert 2.2 \rvert$	371
	6n,7s	4	$\lvert 2.2 \rvert$	
	3x,5x	4	$\lvert 2.4 \rvert$	371
	3n,7a	4	ca. 0.	
	3x,5x	4	ca. $\lvert 2. \rvert$	371
	6n,7s	4	$\lvert 2.4 \rvert$	371

[a] W coupling ($^4J_{HH}$) has been shown to be positive in saturated systems.[471,472]
[b] Observed by deuterium decoupling.[341]
[c] See ref. 342 for estimates of probable errors in these $^nJ_{HH}$ values.
[d] The absolute sign of these allylic couplings ($^4J_{2,4}$) has been shown to be positive.[352,357,481,482]

existence of a stereospecific, four-bond proton–proton coupling. These investigators suggested that this long-range coupling was transmitted via overlap of
the rear orbital lobes at C(5) and C(6) associated with the C(5)—H(5n) and
C(6)—H(6n) bonds, respectively, in **138** [cf. the C(1)–C(3) interaction suggested
by dotted lines in structure **139**. This suggestion derives support from the

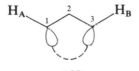

139

observation by Padwa and co-workers[485] that $^4J_{13}$ = ca. 18 Hz for long-range coupling between the bridgehead protons in 2-hydroxy-2-phenylbicyclo[1.1.1]-pentane (**140**). (Bridgehead–bridgehead $^4J_{HH}$ couplings on the order of 18 Hz

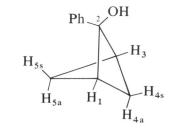

140 ($^4J_{13}$ = ca. 18 Hz; $^4J_{4s5s}$ = 10.0 Hz)[485]

have been observed in other bicyclo[1.1.1]pentane systems as well).[486,487] The now familiar "W coupling" for $^4J_{HH}$ in systems possessing the stereochemistry depicted in **139** has been utilized extensively in conformational analysis; many examples have been reported in rigid bicyclic systems (e.g., substituted norbornanes, norbornenes, bicyclo[2.2.2]octanes, and bicyclo[2.2.2]oct-2-enes; see Table 37).* [371]

In addition to the W-coupling mechanism for $^4J_{HH}$ depicted in structure **139**, some evidence has accrued suggesting that other direct and indirect coupling mechanisms may operate in the transmission of long-range proton–proton couplings in suitably constructed rigid bicyclic systems. All trans propanic coupling (i.e., $^4J_{HH}$ W coupling) has been shown experimentally to be positive in sign.[471,472] However, the results of CNDO/INDO calculations suggest that mutually coupled protons in close spatial proximity to one another can interact via a "through space" (direct) mechanism, the sign of which is negative and the absolute magnitude of which is greater the closer the mutually coupled protons reside.[357,489,490] A third mechanism for long-range proton–proton couplings, suggested by Anet and co-workers,[491] is thought to occur in suitably constructed systems wherein an intervening substituent containing one or more unshared electron pairs ($-\ddot{X}$) "assists" in the transmission of coupling between to distant protons (e.g., $^6J_{AB}$ in half-cage structures **141** and **142**).

Barfield and co-workers[357,466] also have recognized the potential contribution of through-bond (indirect) and other through-space mechanisms for transmission of $^4J_{HH}$ in propanic fragments. They concluded[357] that the interaction depicted in **139** was the dominant (positive, direct) mechanism in this regard which overcomes the negative contribution provided by a competing indirect mechanism in this system.

---

* The importance of the "W arrangement" in the transmission of (conformation-dependent) long-range stereoelectronic effects has been confirmed recently via an *ab initio* study of β-halo-anions in rigid bicyclic systems.[467] The observation of long-range "W-coupling" between protons on the geminal methyl groups in a series of bornane derivatives (for which $^4J_{HH}$ = ca. 0.6–1.2 Hz, CDCl$_3$ solvent) has been utilized for spectral assignment of the methyl group resonances in these compounds.[488]

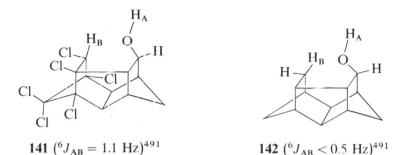

141 $(^6J_{AB} = 1.1$ Hz$)^{491}$ **142** $(^6J_{AB} < 0.5$ Hz$)^{491}$

Several investigators have reported the existence of long-range proton–proton couplings in rigid bicyclic and polycyclic systems that appear to be transmitted over an "extended W" pathway [e.g., all-transoid $^5J_{HH}$ (butanic) and $^6J_{HH}$ (pentanic) systems, as exemplified in structures **143**–**146**]. However,

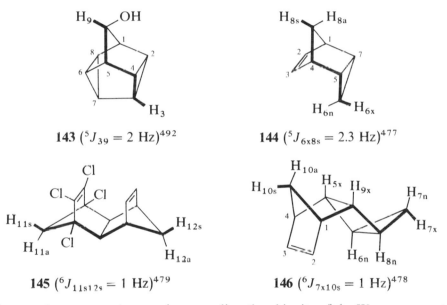

143 $(^5J_{39} = 2$ Hz$)^{492}$ **144** $(^5J_{6x8s} = 2.3$ Hz$)^{477}$

145 $(^6J_{11s12s} = 1$ Hz$)^{479}$ **146** $(^6J_{7x10s} = 1$ Hz$)^{478}$

lest one become overly sanguine regarding the ubiquity of the W arrangement in $^nJ_{HH}$ couplings ($n = 4, 5, 6$), Mark[473] has uttered an appropriate caveat. Substantial coupling (0.85–1.40 Hz) has been observed for $^4J_{6x8a}$ in **147**; the mutually coupled protons adopt a "sickle" configuration (H⌐⌐ᴴ) in this series of compounds. Interestingly, no corresponding "sickle" coupling $^4J_{6n8x}$ could be observed in **147a**–**f**. From these observations, Mark concluded that "as in the case of the W-coupling, a direct interaction between the … coupled nuclei might be operational" (in the coupled sickle arrangement).[473] In view of these results, Mark cautioned that "the presence of significant $^4J_{HH}$ coupling alone cannot serve unequivocally to assign the W-arrangement to protons separated by two carbon–carbon sigma bonds in the norbornane-norbornene skeleton."[473]

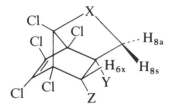

147a $X = O$, $Y = H_{5n}$, $Z = H_{6n}$
147b $X = O$, $Y = Cl$, $Z = H_{6n}$
147c $X = O$, $Y = CH_3$, $Z = H_{6n}$
147d $X = O$, $Y = Ph$, $Z = H_{6n}$
147e $X = O$, $Y = H_{5n}$, $Z = H_{6n}$
147f $X = S$, $Y = CH_3$, $Z = H_{6n}$
147g $X = S$, $Y = H_{5n}$, $Z = CH_3$

Additional long-range couplings involving ^1H nuclei have been studied in rigid bicyclic systems; these include $^nJ_{CH}$ (Table 38), $^nJ_{FH}$ (Table 39), and $^nJ_{PH}$ (Table 40). Of these three $^nJ_{XH}$ couplings, it is the angular dependence of $^nJ_{FH}$ that has been the most extensively studied in appropriately substituted rigid bicyclic (norbornyl and bicyclo[2.2.2]octyl) systems. In their study of long-range proton–fluorine couplings in a series of 5-substituted 7,7-difluoro-1,2,3,4-tetrachloronorbornenes, Williamson and Fenstermaker found that $^4J_{FH}$ values in this system display both positive and negative signs, "being largest when the nuclei conform most closely to the coplanar W-conformation."[339] However, corresponding studies of $^5J_{FH}$ couplings suggest that both orientation and proximity effects must be considered when attempting to analyze long-range proton–fluorine coupling constants. For example, coupling of the ^{19}F nucleus in *syn*-3-fluoro-*anti*-3-bromo-*exo*-tricyclo[3.2.1.0^{2,4}]octane (**148**) to

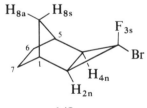

148

TABLE 38. $^nJ_{CH}$ VALUES IN RIGID BICYCLIC RING SYSTEMS RELATED TO NORBORNANE AND BICYCLO[2.2.2]OCTANE

Compound	Carbon Atom Position	Proton Position	Value of n	$^nJ_{CH}$ (Hz)	Reference
	4	1	4	$\lvert 8.7 \rvert$	383

TABLE 39. $^nJ_{FH}$ Values in Rigid Bicyclic Ring Systems Related to Norbornane and Bicyclo[2.2.2]octane

Compound	Fluorine Atom Position	Proton Position	Value of n	$^nJ_{FH}$ (Hz)	Reference
X = OAc	7a	5x	4	−3.24	339
	7a	6x	4	−3.46	
	7a	6n	4	−1.89	
	7s	5x	4	+0.56	
	7s	6x	4	−0.76	
	7s	6n	4	−5.44	
X = OH	7a	5x	4	−3.21	339
	7a	6x	4	−3.48	
	7a	6n	4	−1.95	
	7s	5x	4	+0.36	
	7s	6x	4	−0.45	
	7s	6n	4	−5.38	
X = Cl	7a	5x	4	−3.56	339
	7a	6x	4	−3.21	
	7a	6n	4	−1.56	
	7s	5x	4	+0.81	
	7s	6x	4	−0.68	
	7s	6n	4	−5.27	
X = Ph	7a	5x	4	−3.04	339
	7a	6x	4	−2.51	
	7a	6n	4	−0.90	
	7s	5x	4	+1.04	
	7s	6x	4	+0.39	
	7s	6n	4	−4.95	
X = CO$_2$H	7a	5x	4	−2.81	339
	7a	6x	4	−2.72	
	7a	6n	4	−1.42	
	7s	5x	4	+0.93	
	7s	6x	4	+0.89	
	7s	6n	4	−5.14	
X = CN	7a	5x	4	−2.32	339
	7a	6x	4	−2.74	
	7a	6n	4	−1.40	
	7s	5x	4	+1.12	
	7s	6x	4	−0.32	
	7s	6n	4	−4.87	

TABLE 39 (*continued*)

TABLE 39 (*continued*)

Compound	Fluorine Atom Position	Proton Position	Value of n	$^nJ_{FH}$ (Hz)	Reference
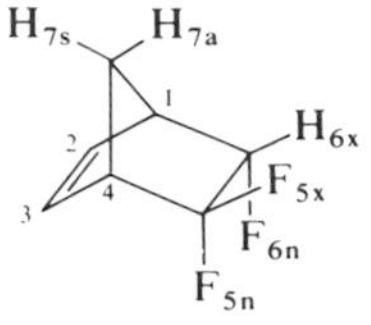	6n	7s	4	$\lvert 6.5 \rvert$	363
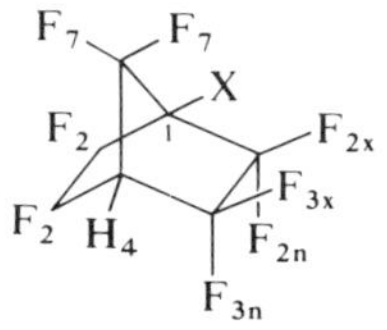					
X = CO$_2$H	2x	4	4	$\lvert 4.4 \rvert$	378
X = I	2x	4	4	$\lvert 4.0 \rvert$	378
X = F	2x	4	4	$\lvert 5.2 \rvert$	378
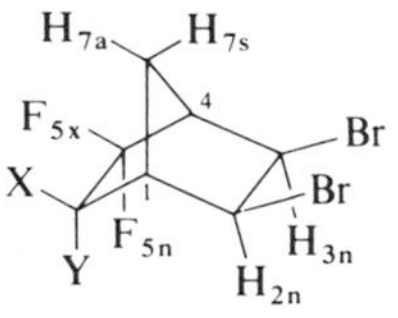					
X = F$_{6x}$, Y = F$_{6n}$	5n	7s	4	$\lvert 5.3 \rvert$	363
X = Y = Cl	5n	7s	4	$\lvert 4.9 \rvert$	363
X = H$_{6x}$, Y = F$_{6n}$	5n	7s	4	$\lvert 7. \rvert$	363
	6n	7a	4	ca. $\lvert 8. \rvert$	
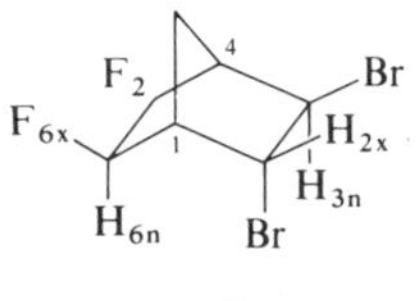	6x	2x	4	ca. $\lvert 3.8 \rvert$	363
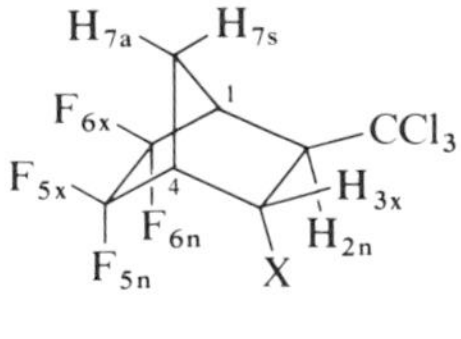 X = Cl	5x	3x	4	$\lvert 3.7 \rvert$	364
	5n	7s	4	$\lvert 5.7 \rvert$	
	6n	7s	4	$\lvert 5.7 \rvert$	

TABLE 39 (*continued*)

TABLE 39 (*continued*)

Compound	Fluorine Atom Position	Proton Position	Value of n	$^nJ_{FH}$ (Hz)	Reference
X = Br	5x	3x	4	ca. $\|4.\|$	364
	5n	7s	4	$\|5.7\|$	
	6n	7s	4	$\|5.7\|$	

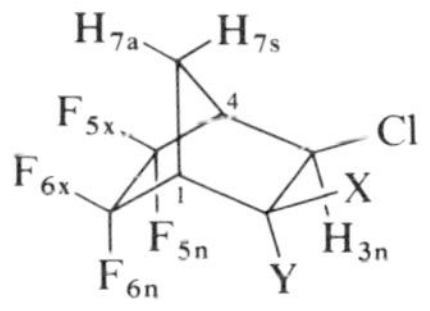					
X = Cl, Y = H_{2n}	5n	7s	4	ca. $\|5.\|$	365
	6n	7s	4	ca. $\|5.\|$	
X = H_{2x}, Y = Cl	5n	7s	4	ca. $\|4.\|$	365
	6n	7s	4	$\|3.-4.\|$	
X = Y = Cl	6n	7s	4	$\|5.8\|$	365

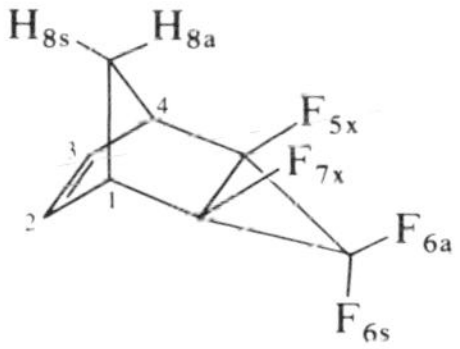	5x	8a	4	$\|6.3\|$	359

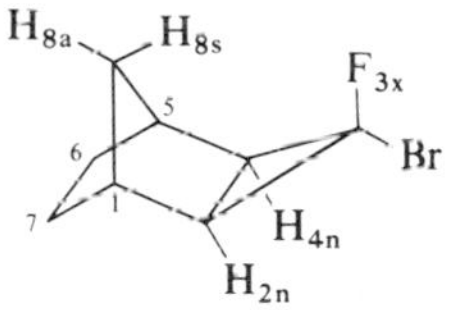	3n	8s	5	$\|3.6\|$	493
	3n	8a	5	$\|3.0\|$	

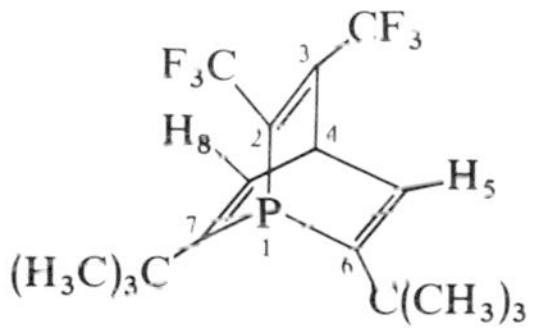	$(CF_3)_3$	5,8	5	$\|2.7\|$	446

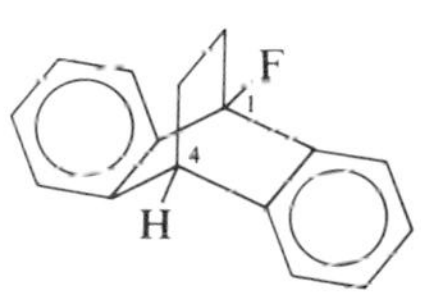	1	4	5	$< \|0.5\|$	494

TABLE 39 (*continued*)

TABLE 39 (*continued*)

Compound	Fluorine Atom Position	Proton Position	Value of n	$^{n}J_{FH}$ (Hz)	Reference
	1	4	5	$< |0.5|$	494
	1	4	5	$|0.88|$	494
	3	8	5	$|2.0|$	495
	3	8	5	$|1.8|$	495

both H_{8s} and to H_{8a} have been observed to be quite similar in magnitude: $(^{5}J_{F_{3s}H\,8s} = 3.6$ Hz, $^{5}J_{F_{3s}H\,8a} = 3.0$ Hz).[493] Small $^{5}J_{FH}$ couplings (all <1 Hz) between the (trans coplanar) $^{19}F_1$ and $^1H_4$ nuclei in systems **149–151** have been reported.[494]

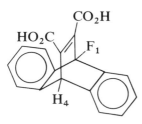

149 $(^{5}J_{F_1H_4} = 0.88 \pm 0.05$ Hz)[494]

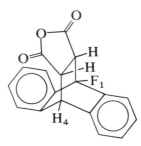

150 (0.3 Hz $< {}^{5}J_{F_1H_4} < 0.5$ Hz)[494]

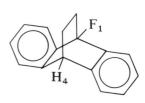

151 (0.3 Hz $< {}^{5}J_{F_1H_4} < 0.5$ Hz)[494]

TABLE 40. $^{n}J_{PH}$ VALUES IN RIGID BICYCLIC RING SYSTEMS RELATED TO NORBORNANE AND BICYCLO[2.2.2]OCTANE

Compound	Phosphorus Atom Position	Proton Position	Value of n	$^{n}J_{PH}$ (Hz)	Reference
H_{7a}, Br, OH, P(OCH$_3$)$_2$, O	2n	7a	4	|5.|	496
H_7, Cl, Cl, Cl, H_{5x}, Cl, Cl, P(OCH$_3$)$_2$, O	5n	7s	4	|4.0|	379

The results of a theoretical study by Wasylishen and Barfield[440] have confirmed the importance of both orientation and proximity effects in determining the sign and magnitude of long-range proton–fluorine coupling constants. Interestingly, long-range $^{n}J_{FH}$ are predicted to be negative when the mutually coupled ^{1}H and ^{19}F nuclei are spatially proximate, whereas large positive values are predicted for, e.g., the particular "sickle" configuration in which the fluorine atom interacts ("direct" mechanism) with the rear lobe of the C(3)—H bond as depicted in structure **152**. In nonrigid systems, relatively small $^{n}J_{FH}$

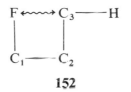

152

couplings are often observed as a result of conformational averaging (which averages arithmetically the individual contributions of sizeable $^{n}J_{FH}$ couplings having opposite signs).[440]

Relatively few long-range couplings involving ^{13}C nuclei have been studied in rigid bicyclic systems. Examples in this regard include $^{n}J_{CC}$ (Table 41), $^{n}J_{FC}$ (Table 42), and $^{n}J_{PC}$ (Table 43).

The conformational dependence of long-range $^{19}F–^{19}F$ couplings is the subject of an extensive theoretical study by Hirao and co-workers.[387] These investigators concluded that the angular dependence of long-range $^{n}J_{FF}$ couplings originated principally from the contribution of the Fermi contact term to the overall coupling constant. Like $^{4}J_{FH}$ couplings, long-range $^{4}J_{FF}$ couplings

TABLE 41. $^nJ_{CC}$ Values in Rigid Bicyclic Ring Systems Related to Norbornane and Bicyclo[2.2.2]octane

Compound[a]	Carbon Atom Positions	Value of n	$^nJ_{CC}$ (Hz)	Reference
97	5,8	4	ca. $\|0.6\|$	286
98	5,8	4	ca. $\|0.3\|$	286
	5,9	4	$\|0.33\|$	289
	9,10	4	$\|0.73\|$	
	9,11	4	$< \|0.11\|$	
	5,9	4	$< \|0.15\|$	289
	9,10	4	$< \|0.11\|$	
	9,11	4	$\|0.33\|$	
	5,8	4	$< \|0.35\|$	285
X = $(\overset{*}{C}O_2H)_{2x}$	5,9	4	$< \|0.5\|$	289
X = $(\overset{*}{C}H_2OH)_{2x}$	5,9	4	$< \|0.2\|$	289

TABLE 41 (*continued*)

TABLE 41 (*continued*)

Compound[a]	Carbon Atom Positions	Value of n	$^nJ_{CC}$ (Hz)	Reference
X = (ĊO₂H)₂ₙ	5,9	4	< |0.7|	289
X = (ĊH₂OH)₂ₙ	5,9	4	< |0.2|	289
	5,9	4	< |0.18|	289
	9,10	4	|0.31|	
	9,11	4	< |0.11|	
	5,9	4	< |0.2|	289
	5,9	4	< |0.4|	285
	5,9	4	|0.42|	289
	9,10	4	< |0.11|	
	9,11	4	< |0.17|	
	5,9	4	|0.29 ± 0.04|	289
	9,10	4	< |0.15|	
	9,11	4	|0.18 ± 0.04|	
X = (ĊO₂H)₁	4,9	4	|0.29 ± 0.07|	289,383
X = (ĊH₂OH)₁	4,9	4	|0.22 ± 0.04|	289,383

[a] Asterisk indicates position of specific ^{13}C labeling.

TABLE 42. $^nJ_{FC}$ Values in Rigid Bicyclic Ring Systems Related to Norbornane and Bicyclo[2.2.2]octane

Compound	Fluorine Atom Position	Carbon Atom Position	Value of n	$^nJ_{FC}$ (Hz)	Reference

Compound	Fluorine Atom Position	Carbon Atom Position	Value of n	$^nJ_{FC}$ (Hz)	Reference
$A = H_{7a}$, $B = H_{7s}$, $W = (CH_3)_{5x}$, $X = H_{6x}$, $Y = H_{5n}$, $Z = H_{6n}$	2x	$(CH_3)_{5x}$	5	$< \lvert 1. \pm 0.5 \rvert$	80
$A = H_{7a}$, $B = H_{7s}$, $W = H_{5x}$, $X = H_{6x}$, $Y = (CH_3)_{5n}$, $Z = H_{6n}$	2x	$(CH_3)_{5n}$	5	$< \lvert 1. \pm 0.5 \rvert$	80
$A = H_{7a}$, $B = H_{7s}$, $W = H_{5x}$, $X = (CH_3)_{6x}$, $Y = H_{5n}$, $Z = H_{6n}$	2x	$(CH_3)_{6x}$	4	ca. $\lvert 1. \pm 0.5 \rvert$	80
$A = H_{7a}$, $B = H_{7s}$, $W = H_{5x}$, $X = H_{6x}$, $Y = H_{5n}$, $Z = (CH_3)_{6n}$	2x	$(CH_3)_{6n}$	4	$< \lvert 1.7 \pm 0.5 \rvert$	80
	2n	$(CH_3)_{6n}$	4	$\lvert 7.0 \pm 0.5 \rvert$	
$A = H_{7a}$, $B = (CH_3)_{7s}$, $W = H_{5x}$, $X = H_{6x}$, $Y = H_{5n}$, $Z = H_{6n}$	2x	$(CH_3)_{7s}$	4	$\lvert 4.5 \pm 0.5 \rvert$	80
	2n	$(CH_3)_{7s}$	4	$< \lvert 1. \pm 0.5 \rvert$	
$A = (CH_3)_{7a}$, $B = H_{7s}$, $W = H_{5x}$, $X = H_{6x}$, $Y = H_{5n}$, $Z = H_{6n}$	2x	$(CH_3)_{7a}$	4	$< \lvert 1. \pm 0.5 \rvert$	80
	1	4	4	$\lvert 3.3 \rvert$	165

TABLE 43. $^nJ_{PC}$ Values in Rigid Bicyclic Ring Systems Related to Norbornane and Bicyclo[2.2.2]octane

Compound	Phosphorus Atom Position	Carbon Atom Position	Value of n	$^nJ_{PC}$ (Hz)	Reference
	8s	9	4	$\lvert 3. \rvert$	307

also display marked stereochemical dependence; positive couplings are predicted for the W arrangement, whereas negative $^4J_{FF}$ values should result from couplings between ^{19}F nuclei in close mutual proximity. Long-range couplings between appropriately situated ^{19}F nuclei can be quite large: For example, in *endo*-di(hexafluoro)cyclopentadiene (**153**), $^5J_{5n8} = |43|$ Hz, $^4J_{2x10a} = |37$ or

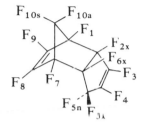

153

$24|$ Hz, and $^4J_{6x10a} = |24$ or $37|$ Hz.[386] Additionally, $^nJ_{FF}$ couplings have been measured in a variety of highly fluorinated norbornanes and norbornenes;[378] representative examples in this regard appear in Table 44.

No reports of long-range ^nJ couplings involving ^{14}N or ^{15}N nuclei in rigid bicyclic systems were found. The only examples of long-range couplings in rigid bicyclic systems of ^1H nuclei with metals appear to involve $^{203/205}Tl$ nuclei; these are presented in Table 45.

TABLE 44. $^nJ_{FF}$ Values in Rigid Bicyclic Ring Systems Related to Norbornane and Bicyclo[2.2.2]octane

Compound	Fluorine Atom Position	Value of n	$^nJ_{FF}$ (Hz)	Reference
X = Y = Br	5x,7a	4	$\|23.6\|$	378
X = Y = I	5x,7a	4	$\|20.2\|$	378
X = Br, Y = H$_4$	2,5x	5	$\|7.1\|$	378
	3,6x	5	$\|5.9\|$	
	5x,7a	4	$\|22.7\|$	
	6x,7a	4	$\|23.2\|$	
X = I, Y = H$_4$	2,5x	5	$\|5.8\|$	378
	3,6x	5	$\|4.9\|$	
	5x,7a	4	$\|21.6\|$	
	6x,7a	4	$\|22.0\|$	
X = H$_1$, Y = H$_4$	5x,7a	4	$\|18.8\|$	378
X = F, Y = Br	2,5n	5	$\|6.6\|$	378
	3,6x	5	$\|6.2\|$	
	6x,7a	4	$\|19.1\|$	
	5x,7a	4	$\|23.2\|$	

TABLE 44 (*continued*)

TABLE 44 (*continued*)

Compound	Fluorine Atom Position	Value of n	$^nJ_{FF}$ (Hz)	Reference
X = F, Y = I	2,5x	5	|7.2|	378
	5x,7a	4	|21.6|	
X = F, Y = H_4	5x,7a	4	|22.1|	378
X = CH_3, Y = I	3,6x	5	|5.3|	378
	3,7a	4	|1.9|	
	2,5x	5	|6.0|	
	2,7a	4	|5.0|	
	6x,7s	4	|0.1|	
	6x,7a	4	|22.8|	378
	6n,7s	4	|4.5|	
	6n,7a	4	|4.3|	
	5x,7s	4	|0.1|	
	5x,7a	4	|21.8|	
	5n,7s	4	|2.0|	
	5n,7a	4	|4.5|	
X = CH_3, Y = H_4	5x,7a	4	|21.0|	378
X = F, Y = CH_3	5x,7a	4	|17.9|	378
X = CF_3, Y = Br	5x,7a	4	|22.2|	378
X = CF_3, Y = I	2,6x	4	|6.8|	378
	5x,7a	4	|22.6|	
X = CF_3, Y = H_4	5x,7a	4	|22.0|	378
X = Y = CH_3	5x,7a	4	|17.9|	378

Compound	Fluorine Atom Position	Value of n	$^nJ_{FF}$ (Hz)	Reference
X = F, Y = H_4, Z = H_3	2,5x	5	|6.0|	378
	6x,7a	4	|19.0|	
	5x,7a	4	|20.5|	
X = Br, Y = F, Z = OCH_3	2,5x	5	|4.9|	378
	2,7a	4	|3.8|	
	5x,7s	4	0.0	
	5x,7a	4	|20.9|	
	5n,7s	4	|2.4|	
	5n,7a	4	|4.5|	
	6x,7s	4	0.0	
	6x,7a	4	|23.5|	
	6n,7s	4	|1.0|	
	6n,7a	4	|5.0|	
X = I, Y = F, Z = OCH_3	2,5x	5	|6.3|	378
	2,7a	4	|2.1|	
	5x,7s	4	0.0	
	5x,7a	4	|19.3|	
	5n,7s	4	|4.2|	
	5n,7a	4	|6.5|	
	6x,7s	4	|0.1|	
	6x,7a	4	|21.9|	
	6n,7s	4	|2.3|	
	6n,7a	4	|4.2|	
X = F, Y = H_4, Z = OCH_3	2,6x	4	|6.5|	378
	6x,7a	4	|22.6|	
	5x,7a	4	|21.6|	

TABLE 44 (*continued*)

TABLE 44 (*continued*)

Compound	Fluorine Atom Position	Value of n	$^{n}J_{FF}$ (Hz)	Reference
$X = I$, $Y = CH_3$, $Z = OCH_3$	2,5x	5	|5.6|	378
	2,7a	4	|3.2|	
	5x,7s	4	|0.8|	
	5x,7a	4	|22.2|	
	5n,7s	4	|6.2|	
	5n,7a	4	|5.0|	
	6x,7s	4	|0.2|	
	6x,7a	4	|21.2|	
	6n,7s	4	|2.2|	
	6n,7a	4	|4.4|	
$X = CF_3$, $Y = I$, $Z = OCH_3$	2,5x	5	|6.6|	378
	6x,7a	4	|17.1|	
	5x,7a	4	|21.2|	

Compound	Fluorine Atom Position	Value of n	$^{n}J_{FF}$ (Hz)	Reference
$R = I$, $W = X = Cl$, $Y = F_{2n}$, $Z = F_{3n}$	3n,5n	4	|81.|	378
	2n,6n	4	|81.|	
	6x,7a	4	|33.|	
	5x,7a	4	|27.|	
$R = W = X = Cl$, $Y = F_{2n}$, $Z = F_{3n}$	3n,5n	4	|81.|	378
	2n,6n	4	|78.|	
	6x,7a	4	|28.|	
	5x,7a	4	|26.|	
$R = H_4$, $W = X = Br$, $Y = F_{2n}$, $Z = F_{3n}$	3n,5n	4	|73.|	378
	2n,6n	4	|76.|	
	6x,7a	4	|30.|	
	5x,7a	4	|31.|	
$R = H_4$, $W = X = Cl$, $Y = F_{2n}$, $Z = F_{3n}$	3n,5n	4	|74.|	378
	2n,6n	4	|75.|	
	6x,7a	4	|29.|	
	5x,7a	4	|25.|	
$R = H_4$, $W = H_{2x}$, $X = H_{3x}$, $Y = F_{2n}$, $Z = F_{3n}$	3n,5n	4	|55.|	378
	2n,6n	4	|54.|	
	6x,7a	4	|17.|	
$R = H_4$, $W = H_{2x}$, $X = F_{3x}$, $Y = F_{2n}$, $Z = H_{3n}$	3x,7s	4	|16.|	378
	2n,6n	4	|47.|	
	6x,7a	4	|20.|	
	5x,7a	4	|17.|	
$R = H_4$, $W = H_{2x}$, $X = OCH_3$, $Y = F_{2n}$, $Z = F_{3n}$	3n,5n	4	|73.|	378
	2n,6n	4	|46.|	
	5x,7a	4	|19.|	
$R = F_4$, $W = OCH_3$, $X = H_{3x}$, $Y = F_{2n}$, $Z = F_{3n}$	3n,5n	4	|45.|	378
	2n,6n	4	|72.|	
	6x,7a	4	|19.|	
	5x,7a	4	|17.|	
$R = CH_3$, $W = H_{2x}$, $X = OCH_3$, $Y = F_{2n}$, $Z = F_{3n}$	3n,5n	4	|81.|	378
	2n,6n	4	|48.|	
	5x,7a	4	|21.|	

TABLE 44 (*continued*)

TABLE 44 (*continued*)

Compound	Fluorine Atom Position	Value of n	$^nJ_{FF}$ (Hz)	Reference
	3n,5n	4	|49.|	378
	2n,6n	4	|84.|	
	6x,7a	4	|20.|	
	5x,7a	4	|20.|	
	8,10a	4	|24.|	386
	9,10a	4	|37.|	
	3,5s	5	|43.8|	

TABLE 45 Miscellaneous $^nJ_{XY}$ Values in Rigid Bicyclic Ring Systems Related to Norbornane and Bicyclo[2.2.2]octane

Compound	X Atom Position	Y Atom Position	Value of n	$^nJ_{XY}$ (Hz)	Reference
	$^{205}Tl_{2x}$	H_4	4	|263.|	50
	$^{205}Tl_{2x}$	H_{6x}	4	|817.|	
X = H_{3x}, Y = H_{3n}	$^{203/205}Tl_{5x}$	H_{3x}	4	|208.|	389
X = H_{3x}, Y = CO_2H	$^{203/205}Tl_{5x}$	H_{3x}	4	|223.|	389

REFERENCES

1. Bartlett, P. D. "Nonclassical Ions: Reprints and Commentary": W. A. Benjamin, Inc.: New York, 1965.
2. Sargent, G. D. *Quart. Rev. (London)*, **1966**, *20*, 301.
3. Bethell, D.; Gold, V. "Carbonium Ions, an Introduction"; Academic Press, Inc.: New York, 1967.
4. Marchand, A. P. Ph.D. Dissertation, University of Chicago, 1965, and references cited therein.
5. Wells, P. R. "Linear Free Energy Relationships"; Academic Press, Inc.: New York, 1968, pp. 43–44.
6. Laszlo, P. *Progr. Nucl. Magn. Resonance Spectrosc.* **1967**, *3*, 231.
7. Engler, E. M.; Laszlo, P. *J. Am. Chem. Soc.* **1971**, *93*, 1317.
8. Marchand, A. P.; Weimar, W. R., Jr.; Segre, A. L.; Ihrig, A. M. *J. Magn. Resonance* **1972**, *6*, 316.
9. Hinckley, C. C. *J. Am. Chem. Soc.* **1969**, *91*, 5160.
10. Mayo, B. C. *Chem. Soc. Rev.* **1973**, *2*, 49.
11. Cockerill, A. F., Davies, G. L. O.; Harden, R. C.; Rackham, D. M. *Chem. Rev.* **1973**, *73*, 553.
12. Sievers, R. E., (Ed.). "Nuclear Magnetic Shift Reagents"; Academic Press, Inc.: New York, 1973.
13. Slonim, I. Y., Bulai, A. K. *Russ. Chem. Rev.* **1973**, *42*, 904.
14. Reuben, J. *Progr. Nucl. Magn. Resonance Spectrosc.* **1973**, *9*, 1.
15. Hofer, O. *Top. Stereochem.* **1976**, *9*, 111.
16. LIS-NMR data have been used to assign stereochemistry to the four isomeric 5-hydroxy-6-methylbicyclo[2.2.2]oct-2-enes: see Willcott, M. R., III; Davis, R. E.: Holder, R. W. *J. Org. Chem.* **1975**, *40*, 1952.
17. See Paasivirta, J. *Suom. Kem.* **1973**, *46B*, 159, 162 and references cited therein.
18. Inagaki, F.; Miyazawa, T. *Progr. NMR Spectrosc.* **1981**, *14*, 67.
19. Marchand, A. P.; Earlywine, A. D.; Goodin, D. B.; Hossain, M. B.; van der Helm, D. "Abstracts of Papers," 178th National Meeting of the American Chemical Society, Washington, D.C., September 10–14, 1979; American Chemical Society: Washington, D.C.; Paper No. ORGN-222.
20. The configurations of other, more complicated dimeric ketones related to system *1* have been elucidated with the aid of LIS-NMR: see Mantzaris, J.; Weissberger, E. *J. Am. Chem. Soc.* **1974**, *96*, 1873.
21. See discussion in Shapiro, B. L.; Johnston, M. D. Jr. *J. Am. Chem. Soc.* **1972**, *94*, 8185.
22. Paasivirta, J.; Hakli, H.; Forsen, K. *Finn. Chem. Lett.* **1974**, 165.

207

23. Hawkes, G. E.; Leibfritz, D.; Roberts, D. W.; Roberts, J. D. *J. Am. Chem. Soc.* **1973**, *95*, 1659.
24. Hawkes, G. E.; Marzin, C.; Johns, S. R.; Roberts, J. D. *J. Am. Chem. Soc.* **1973**, *95*, 1661.
25. However, see Abraham, R.; Chadwick, D. J.; Griffiths, L.; Sancassan, F. *Tetrahedron Lett.* **1979**, 4691.
26. Briggs, J.; Hart, F. A.; Moss, G. P. *Chem. Commun.* **1970**, 1506.
27. Raber, D. J.; Janks, C. M.; Raber, N. K. "Abstracts of Papers," 178th National Meeting of the American Chemical Society, Washington, D.C.; September 10–14, 1979; American Chemical Society, Washington, D.C.; Paper No. ORGN-223.
28. Whitesides, G. M.; Lewis, D. W. *J. Am. Chem. Soc.* **1970**, *92*, 6979.
29. Whitesides, G. M.; Lewis, D. W. *J. Am. Chem. Soc.* **1971**, *93*, 5914.
30. Goering, H. L.; Eikenberry, J. N.; Koermen, G. S. *J. Am. Chem. Soc.* **1971**, *93*, 5913.
31. Kutal, C. In "NMR Shift Reagents," Sievers, R. E., Ed.; Academic Press, Inc.: New York, 1973, pp. 87–98.
32. Sullivan, G. R. *Top. Stereochem.* **1978**, *10*, 287.
33. Kainosho, M.; Ajisaka, K.; Pirkle, W. H.; Beare, S. D. *J. Am. Chem. Soc.* **1972**, *94*, 5924.
34. McKinney, J. D.; Matthews, H. B.; Wilson, N. K. *Tetrahedron Lett.* **1973**, 1895.
35. Noggle, J. H.; Schirmer, R. E. "The Nuclear Overhauser Effect: Chemical Applications"; Academic Press, Inc.: New York, 1971.
36. Bell, R. A.; Saunders, J. K. *Top. Stereochem.* **1973**, *7*, 1.
37. Liu, C.-Y. *Hua Hsueh Tung Pao*, **1978**, *83*; *Chem. Abstr.* **1978**, *89*, 89779u.
38. Backers, G. C.; Schaefer, T. *Chem. Rev.* **1971**, *71*, 616.
39. McFarlane, W.; Rycroft, D. S. *Ann. Rep. NMR Spectrosc.* **1979**, *9*, 319.
40. Saunders, J. K.; Easton, J. W. In "Determination of Organic Structures by Physical Methods," Nachod, F. C.; Zuckerman, J. J.; Randall, E. W. (Eds.); Academic Press, Inc.: New York, 1976, Vol. 6, Chapter 5, pp. 271–333.
41. Schirmer, R. E.; Noggle, J. H.; Davis, J. P.; Hart, P. A. *J. Am. Chem. Soc.* **1970**, *92*, 3266.
42. Anet, F. A. L.; Bourn, A. J. R. *J. Am. Chem. Soc.* **1965**, *87*, 5250.
43. Nouls, J. C.; Van Binst, G.; Martin, R. H. *Tetrahedron Lett.* **1967**, 4065.
44. Scheinmann, F.; Barraclough, D.; Oakland, J. S. *Chem. Commun.* **1970**, 1544.
45. Barraclough, D.; Oakland, J. S.; Scheinmann, F. *J. Chem. Soc. Perkin I*, **1972**, 1500.
46. Farnum, D. G.; Wolf, A. D. *J. Am. Chem. Soc.* **1974**, *96*, 5166.
47. Baldeschwieler, J. D.; Randall, E. W. *Chem. Rev.* **1963**, *63*, 81.
48. Hoffman, R. A.; Forsén, S. *Progr. Magn. Resonance*, **1966**, *1*, 15.
49. Kuhlman, K.; Baldeschwieler, J. D. *J. Am. Chem. Soc.* **1963**, *85*, 1010.
50. Anet, F. A. L. *Tetrahedron Lett.* **1964**, 3399.
51. Hammett, L. P. *J. Am. Chem. Soc.* **1937**, *59*, 96.
52. Hammett, L. P. "Physical Organic Chemistry," 2nd ed.; McGraw-Hill Book Co., Inc.: New York, 1970.
53. Taft, R. W., Jr. In "Steric Effects in Organic Chemistry," Newman, M. S., Ed.; John Wiley and Sons, Inc.: New York, 1956, pp. 556–675.
54. Shorter, J. In "Advances in Linear Free Energy Relationships," Chapman, N. B.; Shorter, J., Eds.; Plenum Press: London, 1972, pp. 71–118.
55. Exner, O. In "Advances in Linear Free Energy Relationships," Chapman, N. B.; Shorter, J., Eds.; Plenum Press: London, 1972, pp. 1–69.
56. Topsom, R. D. *Progr. Phys. Org. Chem.* **1976**, *12*, 1.
57. See Unger, S. H.; Hansch, C. *Progr. Phys. Org. Chem.* **1976**, *12*, 91.
58. For a recent review covering linear correlations of substituent effects in ^{1}H, ^{19}F, and ^{13}C NMR, see Tribble, M. T.; Traynham, J. G. In "Advances in Linear Free Energy Relationships," Chapman, N. B.; Shorter, J., Eds.; Plenum Press: London, 1972, pp. 143–201.
59. Saika, A.; Slichter, C. P. *J. Chem. Phys.* **1954**, *22*, 26.
60. Zürcher, R. F. In "Nuclear Magnetic Resonance in Chemistry," Pesce, B., Ed.; Academic Press, Inc.: New York, 1965, pp. 45–51.
61. Zürcher, R. F. *Progr. Nucl. Magn. Resonance Spectrosc.* **1967**, *2*, 205.
62. Winstein, S.; Carter, P.; Anet, F. A. L.; Bourn, A. J. R. *J. Am. Chem. Soc.* **1965**, *87*, 5247.
63. Marchand, A. P.; Rose, J. E. *J. Am. Chem. Soc.* **1968**, *90*, 3724: Erratum: Marchand, A. P.; Rose, J. E. *J. Am. Chem. Soc.* **1970**, *92*, 5290.
64. Haywood-Farmer, J.; Malkus, H.; Battiste, M. A. *J. Am. Chem. Soc.* **1972**, *94*, 2209.
65. Bukowski, J. A.; Cisak, A. *Rocz. Chem.* **1968**, *42*, 1339.

66. Keith, L. H.; Alford, A. L.; McKinney, J. D. *Tetrahedron Lett.* **1970**, 2489.
67. Parsons, A. M.; Moore, D. J. *J. Chem. Soc.* (*C*), **1966**, 2026.
68. Erdman, J. R.; Simmons, H. E. *J. Org. Chem.* **1968**, *33*, 3808.
69. Foster, R. G.; McIvor, M. C. *J. Chem. Soc.* (*B*), **1969**, 188.
70. McCullough, R.; Rye, A. R.; Wege, D. *Tetrahedron Lett.* **1969**, 5163.
71. Herring, F. G. *Can. J. Chem.* **1970**, *48*, 3498.
72. Mooney, E. F,; Winson, P. H. *Ann. Rev. Nucl. Magn. Resonance Spectrosc.* **1968**, *1*, 243.
73. Emsley, J. W.; Phillips, L. *Progr. Nucl. Magn. Resonance Spectrosc.* **1971**, *7*, 1.
74. Karplus, M.; Das, T. P. *J. Chem. Phys.* **1961**, *34*, 1683.
75. Banks, R. E.; Harrison, A. C.; Haszeldine, R. N.; Orrell, K. G. *J. Chem. Soc.* (*C*), **1967**, 1608.
76. Jefford, C. W.; Kabengele, nT.; Kovacs, J.; Burger, U. *Helv. Chim. Acta* **1974**, *57*, 104.
77. Ando, T.; Yamanaka, H.; Funasaka, W. *Tetrahedron Lett.* **1967**, 2587.
78. Dewar, M. J. S.; Squires, T. G. *J. Am. Chem. Soc.* **1968**, *90*, 210.
79. Jones, K.; Mooney, E. F. *Ann. Rev. Nucl. Magn. Resonance Spectrosc.* **1970**, *3*, 261.
80. Grutzner, J. B.; Jautelat, M.; Dence, J. B.; Smith, R. A.; Roberts, J. D.; *J. Am. Chem. Soc.* **1970**, *92*, 7107.
81. Dence, J. B.; Roberts, J. D. *J. Am. Chem. Soc.* **1969**, *91*, 1542.
82. Gerig, J. T.; Roberts, J. D. *J. Am. Chem. Soc.*, **1966**, *88*, 2791.
83. Lack, R. E.; Roberts, J. D. *J. Am. Chem. Soc.* **1968**, *90*, 6997.
84. Nomura, Y.; Takeuchi, Y. *Chem. Commun.* **1970**, 259.
85. Taft, R. W., Jr. *J. Phys. Chem.* **1960**, *64*, 1805.
86. Taft, R. W.; Price, E.; Fox, I. R.; Lewis, I. C.; Andersen, K. K.; Davis, G. T. *J. Am. Chem. Soc.* **1963**, *85*, 709, 3146.
87. Stothers, J. B. *Quart. Rev.* (*London*), **1965**, *19*, 144.
88. Stothers, J. B. "^{13}C NMR Spectroscopy"; Academic Press, Inc: New York, 1972.
89. Levy, G. C.; Nelson, G. L. "^{13}C NMR for Organic Chemists"; Wiley-Interscience: New York, 1972.
90. Breitmaier, E.; Voelter, W "^{13}C NMR Spectroscopy"; Verlag Chemie GmbH: Weinheim, 1974, 2nd Edition, 1978.
91. Perlin, A. S. In "Isotopes in Organic Chemistry," Buncell, E.; Lee, C. C., Elsevier Publishing Co., Inc.: New York, 1977, Vol. 3, pp. 171–235.
92. Poindexter, G. S.; Kropp, P. J. *J. Org. Chem.* **1976**, *41*, 1215.
93. Lippmaa, E.; Pehk, T.; Paasivirta, J.; Belikova, N.; Platé, A. *Org. Magn. Resonance* **1970**, *2*, 581, and references cited therein.
94. Alger, T. D.; Grant, D. M.; Paul, E. G. *J. Am. Chem. Soc.* **1966**, *88*, 5397.
95. Grant, D. M.; Cheney, B. V. *J. Am. Chem. Soc.* **1967**, *89*, 5315.
96. Cheney, B. V.; Grant, D. M. *J. Am. Chem. Soc.* **1967**, *89*, 5319.
97. Dalling, D. K.; Grant, D. M. *J. Am. Chem. Soc.* **1967**, *89*, 6612.
98. Cheney, B. V. *J. Am. Chem. Soc.* **1968**, *90*, 5386.
99. Forrest, T. P.; Webb, J. G. K. *Org. Magn. Resonance* **1979**, *12*, 371.
100. Schneider, H.-J.; Gschwinder, W.; Weigand, E. F. *J. Am. Chem. Soc.* **1979**, 101, 7195.
101. Stothers, J. B.; Tan, C. T. *Can. J. Chem.* **1976**, *54*, 917.
102. Garratt, P. J.; Riguera, R. *J. Org. Chem.* **1976**, *41*, 465.
103. Brouwer, H.; Stothers, J. B.; Tan, C. T. *Org. Magn. Resonance* **1977**, *9*, 360.
104. Riguera, R. *Tetrahedron*, **1978**, *34*, 2039.
105. Riguera, R.; Garratt, P. J. *An. Quim.* **1978**, *74*, 216; *Chem. Abstr.* **1978**, *89*, 162869e.
106. Stothers, J. B.; Tan, C. T.; Teo, K. C. *Can. J. Chem.* **1973**, *51*, 2893.
107. Grover, S. H.; Guthrie, J. P.; Stothers, J. B.; Tan, C. T. *J. Magn. Resonance* **1973**, *10*, 227.
108. Grover, S. H.; Stothers, J. B. *Can. J. Chem.* **1974**, *52*, 870.
109. Duddeck, H.; Dietrich, W. *Tetrahedron Lett.* **1975**, 2925.
110. Seidman, K.; Maciel, G. E. *J. Am Chem. Soc.* **1977**, *99*, 659.
111. Quin, L. D.; Littlefield, L. B. *J. Org. Chem.* **1978**, *43*, 3508.
112. Littlefield, L. B.; Quin, L. D. *Org. Magn. Resonance* **1979**, *12*, 199.
113. Gorenstein, D. G. *J. Am. Chem. Soc.* **1977**, *99*, 2254.
114. Ehrenson, S.; Brownlee, R. T. C.; Taft, R. W. *Progr. Phys. Org. Chem.* **1973**, *10*, 1.
115. Hehre, W. J.; Taft, R. W.; Topsom, R. D. *Progr. Phys. Org. Chem.* **1976**, *12*, 159.
116. Sichel, J. M.; Whitehead, M. A. *Theor. Chim. Acta* **1966**, *5*, 35.
117. Schaefer, T.; Schneider, W. G. *Can. J. Chem.* **1963**, *41*, 966.
118. Fraser, R. R. *Can. J. Chem.* **1962**, *40*, 78.

119. Musher, J. I. *Mol. Phys.* **1963**, *6*, 93.
120. Paasivirta, J. *Suom. Kem.* **1965**, *38B*, 130.
121. Marchand, A. P.; Marchand, N. W. *Tetrahedron Lett.* **1971**, 1365.
122. Tori, K.; Hata, Y.; Muneyuki, R.; Takano, Y.; Tsuji, T.; Tanida, H. *Can. J. Chem.* **1964**, *42*, 926.
123. Tori, K.; Aono, K.; Hata, Y.; Muneyuki, R.; Tsuji, T.; Tanida, H. *Tetrahedron Lett.* **1966**, 9.
124. Tori, K.; Takano, Y.; Kitahonoki, K. *Chem. Ber.* **1964**, *97*, 2798.
125. Franzus, B.; Baird, W. C., Jr.; Chamberlain, N. F.; Hines, T.; Snyder, E. I. *J. Am. Chem. Soc.* **1968**, *90*, 3721.
126. Inamota, N.; Masuda, S.; Tori, K.; Aono, K.; Tanida, H. *Can. J. Chem.* **1967**, *45*, 1185.
127. Williamson, K. L. *J. Am. Chem. Soc.* **1963**, *85*, 516.
128. Williamson, K. L.; Jacobus, N. C.; Soucy, K. T. *J. Am. Chem. Soc.* **1964**, *86*, 4021.
129. Laszlo, P.; Schleyer, P. von R. *J. Am. Chem. Soc.* **1963**, *85*, 2709.
130. Dailey, B. P.; Shoolery, J. N. *J. Am. Chem. Soc.* **1955**, *79*, 3977.
131. Cavanaugh, J. R.; Dailey, B. P. *J. Chem. Phys.* **1961**, *34*, 1099.
132. Wells, P. R. *Chem. Rev.* **1963**, *63*, 182.
133. Bothner-By, A. A.; Glick, R. E. *J. Chem. Phys.* **1957**, *26*, 1651.
134. Spiesecke, H.; Schneider, W. G. *J. Chem. Phys.* **1961**, *35*, 731.
135. Gurudata; Stothers, J. B. *Can. J. Chem.* **1969**, 47, 3515.
136. Taft, R. W. *J. Am. Chem. Soc.* **1957**, *79*, 1045.
137. Taft, R. W.; Ehrenson, S.; Lewis, I. C.; Glick, R. E. *J. Am. Chem. Soc.* **1959**, *81*, 5352.
138. Taft, R. W.; Prosser, F.; Goodman, L.; Davis, G. T. *J. Chem. Phys.* **1963**, *38*, 380.
139. Wells, P. R.; Ehrenson, S.; Taft, R. W. *Progr. Phys. Org. Chem.* **1968**, *6*, 147.
140. Brownlee, R. T. C.; Taft, R. W. *J. Am. Chem. Soc.* **1970**, *92*, 7007.
141. Fukunaga, J.; Taft, R. W. *J. Am. Chem. Soc.* **1975**, *97*, 1612.
142. Bromilow, J.; Brownlee, R. T. C.; Topsom, R. D.; Taft, R. W. *J. Am. Chem. Soc.* **1976**, *98*, 2020.
143. Dewar, M. J. S.; Marchand, A. P. *J. Am. Chem. Soc.* **1966**, *88*, 3318.
144. Adcock, W.; Dewar, M. J. S. *J. Am. Chem. Soc.* **1967**, *89*, 379.
145. Dewar, M. J. S.; Takeuchi, Y. *J. Am. Chem. Soc.* **1967**, *89*, 391.
146. Adcock, W.; Dewar, M. J. S.; Gupta, B. D. *J. Am. Chem. Soc.* **1973**, *95*, 7353.
147. Adcock, W.; Dewar, M. J. S.; Golden, R.; Zeb, M. A. *J. Am. Chem. Soc.* **1975**, *97*, 2198.
148. Adcock, W.; Alste, J.; Rizvi, S. Q. A.; Aurangzeb, M. *J. Am. Chem. Soc.* **1976**, *98*, 1701.
149. Kitching, W.; Bullpitt, M.; Gartshore, D.; Adcock, W.; Khor, T.-C.; Doddrell, D.; Rae, I. D. *J. Org. Chem.* **1977**, *42*, 2411.
150. Adcock, W.; Cox, D. P. *J. Org. Chem.* **1979**, *44*, 3004.
151. Anderson, G. L.; Parish, R. C.; Stock, L. M.; *J. Am. Chem. Soc.* **1971**, *93*, 6984.
152. Boden, N.; Emsley, J. W.; Feeney, J.; Sutcliffe, L. H. *Mol. Phys.* **1964**, *8*, 133, 467.
153. Dewar, M. J. S.; Grisdale, P. J. *J. Am. Chem. Soc.* **1962**, *84*, 3539, 3541, 3546, 3548.
154. See Cornélis, A.; Lambert, S.; Laszlo, P.; Schaus, P. *J. Org. Chem.* **1981**, *46*, 2130.
155. Reynolds, W. F.; Hamer, G. K. *J. Am. Chem. Soc.* **1976**, *98*, 7296.
156. Anderson, G. L.; Stock, L. M. *J. Am. Chem. Soc.* **1968**, *90*, 212.
157. Anderson, G. L.; Stock, L. M. *J. Am. Chem. Soc.* **1969**, *91*, 6804.
158. Adcock, W.; Khor, T.-C. *J. Org. Chem.* **1977**, *42*, 218.
159. Adcock, W.; Khor, T.-C. *J. Org. Chem.* **1978**, *43*, 1272.
160. Adcock, W.; Khor, T.-C. *J. Am. Chem. Soc.* **1978**, *100*, 7799.
161. Homer, J.; Callaghan, D. *J. Chem. Soc.* (*B*) **1969**, 247.
162. Homer, J.; Callaghan, D. *J. Chem. Soc.* (*B*) **1970**, 1573.
163. Homer, J.; Callaghan, D. *J. Chem. Soc.* (*B*) **1971**, 2430.
164. Adcock, W.; Abeywickrema, A. N. *Tetrahedron Lett.* **1981**, *22*, 1135.
165. Maciel, G. E.; Dorn, H. C. *J. Am. Chem. Soc.* **1971**, *93*, 1268.
166. Wiberg, K. B.; Pratt, W. E.; Bailey, W. F. *Tetrahedron Lett.* **1978**, 4861, 4865.
167. Schneider, H.-J.; Bremser, W. *Tetrahedron Lett.* **1970**, 5197.
168. Grover, S. H.; Marr, D. H.; Stothers, J. B.; Tan, C. T. *Can. J. Chem.* **1975**, *53*, 1351.
169. Stothers, J. B.; Tan, C. T.; Teo, K. C. *Can. J. Chem.* **1973**, *51*, 2893.
170. Ewing, D. F.; Sotheeswaran, S.; Toyne, K. J. *Tetrahedron Lett.* **1977**, 2041.
171. Ewing, D. F.; Toyne, K. J. *J. Chem. Soc., Perkin II*, **1979**, 243.
172. Brown, F. C.; Morris, D. G.; Murray, A. M. *Tetrahedron* **1978**, *34*, 1845.
173. Werstiuk, N. H.; Taillefer, R.; Bell, R. A.; Sayer, B. G.; *Can. J. Chem.* **1972**, *50*, 2146.

174. Lippmaa, E.; Pehk, T.; Belikova, N. A.; Bobyleva, A. A.; Kalinichenko, A. N.; Ordubadi, M. D.; Platé, A. F. *Org. Magn. Resonance* **1976**, *8*, 74.
175. Poindexter, G. S.; Kropp, P. J. *J. Org. Chem.* **1976**, *41*, 1215.
176. Bach, R. D.; Holubka, J. W.; Taaffee, T. H. *J. Org. Chem.* **1979**, *44*, 35.
177. Brouwer, H.; Stothers, J. B.; Tan, C. T. *Org. Magn. Resonance* **1977**, *9*, 360.
178. Reference 89, p. 48.
179. Duddeck, H.; Wolff, P. *Org. Magn. Resonance* **1977**, *9*, 528.
180. Quarroz, D.; Sonney, J.-M.; Chollet, A.; Florcy, A.; Vogel, P. *Org. Magn. Resonance*, **1977**, *9*, 611.
181. Schneider, H.-J.; Freitag, W. *J. Am. Chem. Soc.* **1977**, *99*, 8363.
182. Duddeck, H.; Fenerhelm, H.-T. *Tetrahedron*, **1980**, *36*, 3009.
183. Adcock, W.; Aldous, G. L.; Kitching, W. *J. Organometal. Chem.* **1980**, *202*, 385.
184. Adcock, W.; Aldous, G. L. *J. Organometal. Chem.* **1980**, *201*, 411.
185. Olah, G. A.; Westerman, P. W. *J. Am. Chem. Soc.* **1973**, *95*, 7530.
186. Fraenkel, G.; Farnum, D. G. In "Carbonium Ions," Olah, G. A.; von R. Schleyer, P., Eds.; Wiley-Interscience: New York, 1968, Vol. 1, Chapter 7, pp. 237–255.
187. Farnum, D. G. *Adv. Phys. Org. Chem.* **1975**, *11*, 123.
188. Kramer, G. M. *Adv. Phys. Org. Chem.* **1975**, *11*, 177.
189. Young, R. N. *Progr. NMR Spectrosc.* **1979**, *12*, 261.
190. Winstein, S. *Quart. Rev. (London)*, **1969**, *23*, 1411.
191. Brown, H. C. *Acc. Chem. Res.* **1973**, *6*, 377.
192. Sargent, G. D. In "Carbonium Ions," Olah, G.; Schleyer, P. von R., Eds.; Wiley—Interscience: New York, 1972, Vol. III, Chapter 24, pp. 1099–1200.
193. Brown, H. C.; Schleyer, P. von R. "The Nonclassical Ion Problem"; Plenum Press: New York, 1977.
194. Brown, H. C. *Top. Current Chem.* **1979**, *80*, 1.
195. Hogeveen, H.; van Kruchten, E. M. G. A. *Top. Current Chem.* **1979**, *80*, 89.
196. Kirmse, W. *Top. Current Chem.* **1979**, *80*, 125.
197. Brown, H. C. *Tetrahedron*, **1976**, *32*, 179.
198. Olah, G. A. *Acc. Chem. Res.* **1976**, *9*, 41.
199. Olah, G. A. *Top. Current Chem.* **1979**, *80*, 19.
200. Saunders, M.; Schleyer, P. von R.; Olah, G. A. *J. Am. Chem. Soc.* **1964**, *86*, 5680.
201. Olah, G. A.; White, A. M.; DeMember, J. R.; Commeyras, A.; Lin, C. Y. *J. Am. Chem. Soc.* **1970**, *92*, 4627.
202. Dewar, M. J. S.; Marchand, A. P. *Ann. Rev. Phys. Chem.* **1965**, *16*, 321.
203. Olah, G. A.; Liang, G., Matteescu, G. D.; Riemenschneider, J. L. *J. Am. Chem. Soc.* **1973**, *95*, 8698.
204. Brown, H. C.; Peters, E. N. *J. Am. Chem. Soc.* **1977**, *99*, 1712.
205. Nelson, G. L.; Williams, E. A. *Progr. Phys. Org. Chem.* **1976**, *12*, 229.
206. Olah, G. A.; White, A. M. *J. Am. Chem. Soc.* **1969**, *91*, 3954, 3956.
207. Olah, G. A.; White, A. M. *J. Am. Chem. Soc.* **1969**, *91*, 5801.
208. ^{13}C NMR spectrum of **81**: Olah, G. A.; White, A. M. *J. Am. Chem. Soc.* **1969**, *91*, 6883.
209. Olah, G. A.; Liang, G. *J. Am. Chem. Soc.* **1974**, *96*, 195.
210. Olah, G. A.; DeMember, J. R.; Lui, C. Y.; Porter, R. D. *J. Am. Chem. Soc.* **1971**, *93*, 1442.
211. Saunders, M.; Telkowski, L.; Kates, M. R. *J. Am. Chem. Soc.* **1977**, *99*, 8070.
212. Farnum, D. G.; Wolf, A. D. *J. Am. Chem. Soc.* **1974**, *96*, 5166.
213. Farnum, D. G.; Mehta, G. *J. Am. Chem. Soc.* **1969**, *91*, 3256.
214. Olah, G. A.; Liang, G. *J. Am. Chem. Soc.* **1973**, *95*, 3792.
215. Olah, G. A.; Liang, G. *J. Am. Chem. Soc.* **1975**, *97*, 1920.
216. Olah, G. A.; Liang, G. *J. Am. Chem. Soc.* **1975**, *97*, 6803.
217. Olah, G. A.; Surya Prakash, G. K.; Rawdah, T.-N.; Whittaker, D.; Rees, J. C. *J. Am. Chem. Soc.* **1979**, *101*, 3935.
218. ^{1}H NMR spectrum of **81**: Lustgarten, R. K.; Brookhart, M.; Winstein, S. *J. Am. Chem. Soc.* **1972**, *94*, 2347.
219. ^{1}H NMR spectrum of **81**: Story, P. R.; Saunders, M. *J. Am. Chem. Soc.* **1960**, *82*, 6199.
220. ^{1}H NMR spectrum of **81**: Story, P. R.; Snyder, L. C.; Douglass, D. C.; Anderson, E. W.; Kornegay, R. L. *J. Am. Chem. Soc.* **1963**, *85*, 3630.
221. Olah, G. A.; Liang, G. *J. Am. Chem. Soc.* **1975**, *97*, 2236.
222. Olah, G. A.; Liang, G. *J. Am. Chem. Soc.* **1976**, *98*, 6304.

223. Olah, G. A.; Surya Prakash, G. K.; Rawdah, T. N. *J. Am. Chem. Soc.* **1980**, *102*, 6127.
224. Brown, H. C.; Takeuchi, K.; Ravindranathan, M. *J. Am. Chem. Soc.*, **1977**, *99*, 2684, and references cited therein.
225. Brown, H. C.; Rao, C. G. *J. Org. Chem.* **1979**, *44*, 3536.
226. Farnum, D. G.; Botto, R. E.; Chambers, W. T.; Lam, B. *J. Am. Chem. Soc.* **1978**, *100*, 3847.
227. Gassman, P. G.; Zeller, J.; Lumb, J. T. *Chem. Commun.* **1968**, 69.
228. Gassman, P. G.; Fentiman, A. F., Jr. *J. Am. Chem. Soc.* **1969**, *91*, 1545.
229. Gassman, P. G.; Fentiman, A. F., Jr. *J. Am. Chem. Soc.* **1970**, *92*, 2549.
230. Richey, H. G., Jr.; Nichols, D.; Gassman, P. G.; Fentiman, A. F., Jr.; Winstein, S.; Brookhart, M.; Lustgarten, R. K. *J. Am. Chem. Soc.* **1970**, *92*, 3783.
231. Olah, G. A.; Berrier, A. L.; Arvanaghi, M.; Surya Prakash, G. K. *J. Am. Chem. Soc.* **1981**, *103*, 1122.
232. Olah, G. A.; Surya Prakash, G. K.; Liang, G. *J. Am. Chem. Soc.* **1977**, *99*, 5683.
233. Saunders, M.; Telkowski, L.; Kates, M. R. *J. Am. Chem. Soc.* **1977**, *99*, 8070.
234. Saunders, M.; Kates, M. R. *J. Am. Chem. Soc.* **1977**, *99*, 8071.
235. Saunders, M.; Kates, M. R.; Wiberg, K. B.; Pratt, W. *J. Am. Chem. Soc.* **1977**, *99*, 8072.
236. Saunders, M.; Kates, M. R. *J. Am. Chem. Soc.* **1978**, *100*, 7082.
237. Saunders, M.; Kates, M. R. *J. Am. Chem. Soc.* **1980**, *102*, 6867.
238. Saunders, M.; Chandrasekhar, J.; Schleyer, P. von R. In "Rearrangements in Ground and Excited States," de Mayo, P., Ed.; Academic Press, Inc.: New York, 1980, Vol. 1, Chapter 1.
239. Schleyer, P. von R.; Lenou, D.; Mison, P.; Liang, G.; Surya Prakash, G. K.; Olah, G. A. *J. Am. Chem. Soc.* **1980**, *102*, 683.
240. Williams, R. E.; Field, L. D. *Pure Applied Chem.* in press.
241. Servis, K. L.; Shue, F.-F. *J. Am. Chem. Soc.* **1980**, *102*, 7233.
242. Kelly, D. P.; Farquharson, G. J.; Giansiracusa, J. J.; Jensen, W. A.; Hügel, H. M.; Porter, A. P.; Rainbow, I. J.; Timewell, P. H. *J. Am. Chem. Soc.* **1981**, *103*, 3539.
243. Brown, H. C.; Periasamy, M. *J. Org. Chem.* **1981**, *46*, 3166.
244. Brown, H. C.; Kelly, D. P.; Periasamy, M. *J. Org. Chem.* **1981**, *46*, 3170.
245. McFarlane, W. *Quart. Rev. (London)*, **1969**, *23*, 187.
246. Reference 88, pp. 332–348.
247. Shoolery, J. N. *J. Chem. Phys.* **1959**, *31*, 1427.
248. Muller, N.; Pritchard, D. E. *J. Chem. Phys.* **1959**, *31*, 768, 1471.
249. Muller, N. *J. Chem. Phys.* **1962**, *36*, 359.
250. Ramsey, N. F. *Phys. Rev.* **1953**, *91*, 303.
251. Pople, J. A.; McIver, J. W., Jr.; Ostlund, N. S. *J. Chem. Phys.* **1968**, *49*, 2960, 2965.
252. Maciel, G. E.; McIver, J. W., Jr.; Ostlund, N. S.; Pople, J. A. *J. Am. Chem. Soc.* **1970**, *92*, 1, 11.
253. Ellis, P. D.; Maciel, G. E. *J. Am. Chem. Soc.* **1970**, *92*, 5829.
254. Bartuska, V. J.; Maciel, G. E. *J. Magn. Resonance* **1971**, *5*, 211.
255. Bartuska, V. J.; Maciel, G. E. *J. Magn. Resonance* **1972**, *7*, 36.
256. Summerhays, K. D.; Maciel, G. E. *J. Am. Chem. Soc.* **1972**, *94*, 8348.
257. Christl, M. *Chem. Ber.* **1975**, *108*, 2781.
258. Laszlo, P.; Schleyer, P. von R. *J. Am. Chem. Soc.* **1964**, *86*, 1171.
259. Saupe, A.; Englert, G. *Phys. Rev. Lett.* **1963**, *11*, 462.
260. Bernheim, R. A.; Lavery, B. J. *J. Am. Chem. Soc.* **1967**, *89*, 1279.
261. Buckingham, A. D.; McLauchlan, K. A.; *Proc. Chem. Soc.* **1963**, 144.
262. Mackor, E. L.; McLean, C. *J. Chem. Phys.* **1966**, *44*, 64.
263. Karplus, M. *J. Am. Chem. Soc.* **1962**, *84*, 2458.
264. Tori, K.; Muneyuki, R.; Tanida, H. *Can. J. Chem.* **1963**, *41*, 3142.
265. Tori, K.; Tsushima, T.; Tanida, H.; Kushida, K.; Satoh, S. *Org. Magn. Resonance* **1974**, *6*, 324.
266. Chatterhee, N. *J. Magn. Resonance*, **1979**, *33*, 241.
267. Cooper, M. A.; Manatt, S. L. *Org. Magn. Resonance* **1970**, *2*, 511.
268. Garratt, D. G.; Ryan, M. D.; Beaulieu, P. L. *J. Org. Chem.* **1980**, *45*, 839.
269. White, A. M.; Olah, G. A. *J. Am. Chem. Soc.* **1969**, *91*, 2943.
270. Reference 88, pp. 235–236, 346–348.
271. Olah, G. A.; Yu, S. H. *J. Org. Chem.* **1975**, *40*, 3638.
272. Olah, G. A.; DeMember, J. R.; Lui, C. Y.; White, A. M. *J. Am. Chem. Soc.* **1969**, *91*, 3958.
273. Gil, V. M. S.; Geraldes, C. F. G. C. In "Nuclear Magnetic Resonance Spectroscopy of

Nuclei Other Than Protons," Axenrod, T.; Webb, G. A., Eds.; Wiley-Interscience: New York, 1974, Chapter 14, pp. 219–231.

274. Malinowsky, E. R. *J. Am. Chem. Soc.* **1961**, *83*, 4479.
275. Emsley, J. W.; Phillips, L.; Wray, V. *Progr. NMR Spectrosc.* **1977**, *10*, 85.
276. Tiers, G. V. D. *J. Phys. Chem.* **1963**, *67*, 1373.
277. Della, E. W.; Cotsaris, E.; Hine, P. T. *J. Am. Chem. Soc*, **1981**, *103*, 4131.
278. Maciel, G.; Dorn, H. C., Greene, R. L.; Kleschick, W. A.; Peterson, M. R., Jr.; Wahl, G. H. *Org. Magn. Resonance* **1974**, *6*, 178.
279. Weigert, F.; Roberts, J. D. *J. Am. Chem. Soc.* **1972**, *94*, 6021.
280. Maciel, G. E. In "Nuclear Magnetic Resonance Spectroscopy of Nuclei Other Than Protons," Axenrod, T.; Webb, G. A., Eds.; Wiley-Interscience: New York, 1974, Chapter 13, pp. 187–218.
281. Wray, V.; Ernst, L. cited in ref. 450.
282. McLauchlan, K. A. *Chem. Commun.* **1965**, 105.
283. Dreeskamp, H.; Hildenbrand, K.; Pfisterer, G. *Mol. Phys.* **1968**, *14*, 295.
284. Grant, D. M. *J. Am. Chem. Soc.* **1967**, *89*, 2228.
285. Barfield, M.; Burfitt, I.; Doddrell, D. *J. Am. Chem. Soc.* **1975**, *97*, 2631.
286. Marshall, J. L.; Miiller, D. E. *J. Am. Chem. Soc.*, **1973**, *95*, 8305.
287. Marshall, J. L.; Conn, G. A.; Barfield, M. *Org. Magn. Resonance* **1977**, *9*, 404.
288. Marshall, J. L.; McDaniel, C. R. "Abstracts of Papers,", 35th Southwest Regional Meeting of the American Chemical Society, Austin, Texas, December 5–7, 1979; American Chemical Society: Washington, D.C., 1979.
289. Marshall, J. L. personal communication; see also ref. 286.
290. Marshall, J. L.; Barfield, M. Cited in ref. 419, Table 3-2.
291. Berger, S. *Org. Magn. Resonance* **1980**, *14*, 65.
292. Berger, S. *J. Org. Chem.* **1978**, *43*, 209.
293. Valckx, L. A.; Borremans, F. A. M.; Becu, C. E.; DeWaele, R. H. K.; Anteunis, M. J. O. *Org. Magn. Resonance* **1979**, *12*, 302.
294. Berger, S.; Roberts, J. D. *J. Am. Chem. Soc.* **1974**, *96*, 6757.
295. Lichter, R. L.; Fehder, C. G.; Patton, P. H.; Coombs, J.; Dorman, D. E. *Chem. Commun.* **1974**, 114.
296. Berger, S. *Tetrahedron* **1978**, *34*, 3133.
297. Wasylishen, R. E. *Ann. Rep. NMR Spectrosc.* **1977**, *7*, 245.
298. Lichter, R. L. In "Determination of Organic Structures by Physical Methods," Nachod, F. C.; Zuckerman, J. J., Eds.; Academic Press, Inc.: New York, 1971, Vol. 4, Chapter 4, pp. 196–232.
299. Anteunis, M. J. O.; Borremans, F. A. M.; Gelan, J.; Marchand, A. P.; Allen, R. W. *J. Am. Chem. Soc.* **1978**, *100*, 4050.
300. Mooney, E. F.; Winson, P. H. *Ann. Rev. NMR Spectrosc.* **1969**, *2*, 125.
301. Yeh, H. J. C.; Ziffer, H.; Jerina, D. M.; Boyd, D. R.; *J. Am. Chem. Soc.* **1973**, *95*, 2741.
302. Schulman, J. M.; Venanzi, T. *J. Am. Chem. Soc.* **1976**, *98*, 4701.
303. Schulman, J. M.; Venanzi, T. *J. Am. Chem. Soc.* **1976**, *98*, 6739.
304. Randall, E. W.; Gillies, D. G. *Progr. NMR Spectrosc.* **1970**, *6*, 119.
305. Binsch, G.; Lambert, J. B.; Roberts, B. W.; Roberts, J. D. *J. Am. Chem. Soc.* **1964**, *86*, 5564.
306. Wetzel, R. B.; Kenyon, G. L. *J. Am. Chem. Soc.* **1972**, *94*, 9230.
307. Lauer, M.; Samuel, O.; Kagan, H. B. *J. Organometal. Chem.* **1979**, *177*, 309.
308. Wetzel, R. B.; Kenyon, G. L. *Chem. Commun.* **1973**, 287.
309. Wetzel, R. B.; Kenyon, G. L. *J. Am. Chem. Soc.* **1974**, *96*, 5189.
310. Thiem, J.; Meyer, B. *Org. Magn. Resonance* **1978**, *11*, 50.
311. Buchanan, G. W.; Benezra, C. *Can. J. Chem.* **1976**, *54*, 231.
312. McFarlane, W. *Proc. R. Soc. London Ser. A.* **1968**, *306*, 185.
313. Buchanan, G. W.; Morin, F. G. *Can. J. Chem.*, **1977**, *55*, 2885.
314. Buchanan, G. W.; Bowen, J. H. *Can. J. Chem.* **1977**, *55*, 604.
315. Thiem, J.; Meyer, B. *Tetrahedron Lett.* **1977**, 3573.
316. Doddrell, D.; Burfitt, I.; Kitching, W.; Bullpitt, M.; Lee, C.-H.; Mynott, R. J.; Considine, J. L.; Kuivila, H. G.; Sarma, R. H. *J. Am. Chem. Soc.* **1974**, *96*, 1640.
317. Kitching, W.; Praeger, D.; Doddrell, D.; Anet, F. A. L.; Krane, J. *Tetrahedron Lett.* **1975**, 759.

318. Uemura, S.; Miyoshi, H.; Okano, M.; Morishima, I.; Inubushi, T. *J. Organmetal. Chem.* **1979**, *165*, 9.
319. Morishima, I.; Inubushi, T.; Uemura, S.; Miyoshi, H.; *J. Am. Chem. Soc.* **1978**, *100*, 354.
320. Barron, P. F.; Doddrell, D.; Kitching, W. *J. Organometal Chem.* **1977**, *132*, 351.
321. Uemura, S.; Miyoshi, H.; Okano, M.; Morishima, I.; Inubushi, T. *J. Organometal. Chem.* **1979**, *171*, 131.
322. Nesmeyanov, A. N.; Aleksanyan, V. T.; Denisovich, L. I.; Nekrasov, Yu. S.; Fedin, E. I.; Khvostenko, V. I.; Kritskaya, I. I.; *J. Organometal. Chem.* **1979**, *172*, 133.
323. Bothner-By, A. A. *Adv. Magn. Resonance* **1965**, *1*, 195.
324. Sternhell, S. *Quart. Rev. (London)* **1969**, *23*, 236.
325. Günther, H.; Jikeli, G. *Chem. Rev.* **1977**, *77*, 599.
326. Barfield, M.; Grant, D. M. *J. Am. Chem. Soc.* **1963**, *85*, 1899, and references cited therein.
327. Cox, R.; Smith, S. *J. Phys. Chem.* **1967**, *71*, 1809.
328. Werstiuk, N. H. *Can. J. Chem.* **1970**, *48*, 2310.
329. Subramanian, P. M.; Emerson, M. T.; LeBel, N. A. *J. Org. Chem.* **1965**, *30*, 2624.
330. Kamezawa, N.; Sakashita, K.; Hayamizu, K. *Org. Magn. Resonance* **1969**, *1*, 405.
331. Davis, J. C., Jr.; Van Auken, T. V. *J. Am. Chem. Soc.* **1965**, *87*, 3900.
332. Paasivirta, J. *Suom. Kem.* **1963**, *36B*, 76.
333. Anet, F. A. L.; Lee, H. H.; Sudmeier, J. L. *J. Am. Chem. Soc.* **1967**, *89*, 4431.
334. Gassend, R.; Limouzin, Y.; Maire, J.-C. *Org. Magn. Resonance* **1974**, *6*, 259.
335. Marshall, J. L.; Walter, S. R.; Barfield, M.; Marchand, A. P.; Marchand, N. W.; Segre, A. L. *Tetrahedron* **1976**, *32*, 537.
336. Marshall, J. L.; Seiwell, R. *Org. Magn. Resonance* **1976**, *8*, 419.
337. Cain, A. H.; Roberts, J. D. *J. Org. Chem.* **1977**, *42*, 2853.
338. Benezra, C. *J. Am. Chem. Soc.* **1973**, *95*, 6890.
339. Williamson, K. L.; Fenstermaker, J. C. *J. Am. Chem. Soc.* **1968**, *90*, 342.
340. Marchand, A. P.; Cornell, D. R.; Hopla, R. E.; Fowler, B. N.; Washburn, D. D.; Zinsser, C. C. *Tetrahedron Lett.* **1972**, 3277.
341. Marchand, A. P.; Marchand, N. W.; Segre, A. L. *Tetrahedron Lett.* **1969**, 5207.
342. Marshall, J. L.; Walter, S. R. *J. Am. Chem. Soc.* **1974**, *96*, 6358.
343. Tori, K.; Kitahonoki, K.; Takano, Y.; Tanida, H.; Tsuji, T.; *Tetrahedron Lett.* **1964**, 559.
344. Simmons, H. E. *J. Am. Chem. Soc.* **1961**, *83*, 1657.
345. Callot, H. J.; Benezra, C. *Can. J. Chem.* **1972**, *50*, 1078.
346. Ogoshi, H.; Setsune, J.-I.; Yoshida, Z.-I. *J. Organometal. Chem.* **1980**, *185*, 95.
347. Foster, R. G.; McIvor, M. C. *Org. Magn. Resonance* **1969**, *1*, 203.
348. Cava, M. P.; Scheel, F. M. *J. Org. Chem.* **1967**, *32*, 1304.
349. Katz, T. J.; Carnahan, J. C., Jr.; Boecke, R. *J. Org. Chem.* **1967**, *32*, 1301.
350. Arnold, D. R.; Trecker, D. J.; Whipple, E. B. *J. Am. Chem. Soc.* **1965**, *87*, 2596.
351. Paddon-Row, M. N.; Hartcher, R. *J. Am. Chem. Soc.* **1980**, *102*, 671.
352. Cimarusti, C. M.; Wolinsky, J. *J. Am. Chem. Soc.* **1968**, *90*, 113.
353. Cristol, S. J.; Russell, T. W.; Mohrig, J. R.; Plorde, D. E. *J. Org. Chem.* **1966**, *31*, 581.
354. Williamson, K. L.; Hsu, Y.-F. L.; Lacko, R.; Youn, C. H. *J. Am. Chem. Soc.* **1969**, *91*, 6129.
355. Williamson, K. L.; Hsu, Y.-F. L.; Young, E. I. *Tetrahedron* **1968**, *24*, 6007.
356. Ramey, K. C.; Lini, D. C.; Moriarty, R. M.; Gopal, H.; Welsh, H. G. *J. Am. Chem. Soc.* **1967**, *89*, 2401.
357. Barfield, M.; Dean, A. M.; Fallick, C. J.; Spear, R. J.; Sternhell, S.; Westerman, P. W. *J. Am. Chem. Soc.* **1975**, *97*, 1482.
358. Sargent, P. B. *J. Am. Chem. Soc.* **1969**, *91*, 3061.
359. Chalier, G.; Gagnaire, D.; Rassat, A. *Bull. Soc. Chim. France* **1969**, 387.
360. Martin, M. M.; Koster, R. A. *J. Org. Chem.* **1968**, *33*, 3428.
361. Davies, D. I.; Dowle, M. D.; Kenyon, R. F. *Synthesis* **1979**, 990.
362. Cristol, S. J.; Nachtigall, G. W. *J. Org. Chem.* **1967**, *32*, 3738.
363. Smart, B. E. *J. Org. Chem.* **1973**, *38*, 2027.
364. Smart, B. E. *J. Org. Chem.* **1973**, *38*, 2035.
365. Smart, B. E. *J. Org. Chem.* **1973**, *38*, 2039.
366. Fehnel, E. A.; Brokaw, F. C. *J. Org. Chem.* **1980**, *45*, 578.
367. Yasuda, M.; Harano, K.; Kanematsu, K. *J. Org. Chem.* **1980**, *45*, 659.
368. Rammash, B. K.; Gladstone, C. M.; Wong, J. L. *J. Org. Chem.* **1981**, *46*, 3036.

369. Cardenas, C. G. *J. Org. Chem.* **1971**, *36*, 1631.
370. Anteunis, M. J. O.; Marchand, A. P.; Allen, R. W. Unpublished results.
371. Jung, M. E.; Shapiro, J. J. *J. Am. Chem. Soc.* **1980**, *102*, 7862.
372. Manatt, S. L.; Ellenian, D. D.; Brois, S. J. *J. Am. Chem. Soc.* **1965**, *87*, 2220.
373. Reference 88, p. 353.
374. Sackmann, E.; Dreeskamp, H. *Spectrochim. Acta* **1965**, *21*, 2005.
375. Tan, R. Y. S.; Russell, R. A.; Warrener, R. N. *Tetrahedron Lett.* **1979**, 5031.
376. Williamson, K. L.; Hsu, Y.-F. L.; Hall, F. H.; Swager, S.; Coulter, M. S. *J. Am. Chem. Soc.* **1968**, *90*, 6717.
377. Ihrig, A. M.; Smith, S. L. *J. Am. Chem. Soc.* **1970**, *92*, 759.
378. Lindon, J. C. Ph.D. Dissertation, University of Birmingham, 1969, cited in ref. 275.
379. Callot, H. J.; Benezra, C. *Can. J. Chem.* **1970**, *48*, 3382.
380. Evelyn, L.; Hall, L. D.; Steiner, P. R.; Stokes, D. H. *Org. Magn. Resonance* **1973**, *5*, 141.
381. Coskran, K. J.; Verkade, J. G. *Inorg. Chem.* **1965**, *4*, 1655.
382. Bothner-By, A. A.; Cox, R. H. *J. Phys. Chem.* **1969**, *73*, 1830.
383. Barfield, M.; Brown, S. E.; Canada, E. D., Jr.; Ledford, N. D.; Marshall, J. L.; Yakali, E. **1980**, *J. Am. Chem. Soc.* **1980**, *102*, 3355.
384. Mesch, K. A.; Quin, L. D. *Tetrahedron Lett.* **1980**, *21*, 4791.
385. Sørensen, S.; Hansen, R. S.; Jakobsen, H. J. *J. Am. Chem. Soc.* **1972**, *94*, 5900.
386. Fields, R.; Green, M.; Jones, A. *J. Chem. Soc. (B)* **1967**, 270.
387. Hirao, K.; Natatsuji, H.; Kato, H. *J. Am. Chem. Soc.* **1973**, *95*, 31.
388. Evans, D. F.; Manatt, S. L.; Elleman, D. D.; *J. Am. Chem. Soc.* **1965**, *85*, 238.
389. McKillop, A.; Ford, M. E.; Taylor, E. C. *J. Org. Chem.* **1974**, *39*, 2434.
390. Karplus, M. *J. Chem. Phys.* **1959**, *30*, 11.
391. Karplus, M. *J. Am. Chem. Soc.* **1963**, *85*, 2870.
392. Altona, C.; Buys, H. R.; Hageman, H. J.; Havinga, E. *Tetrahedron* **1967**, *23*, 2265.
393. Bystrov, V. F.; Ivanov, V. T.; Portnova, S. L.; Balashova, T. A.; Ovchinnikov, Y. A. *Tetrahedron* **1973**, *29*, 873.
394. Reynolds, W. F.; Schaefer, T. *Can. J. Chem.* **1964**, *42*, 2119.
395. Bauld, N. L.; Rim, Y. S. *J. Org. Chem.* **1968**, *33*, 1303.
396. Stolow, R. D.; Gallo, A. A. *Tetrahedron Lett.* **1968**, 3331.
397. Kiefer, E. F.; Gericke, W.; Amimoto, S. T. *J. Am. Chem. Soc.* **1968**, *90*, 6246.
398. Williamson, K. L.; Li, Y.-F.; Hall, F. H.; Swager, S. *J. Am. Chem. Soc.* **1966**, *88*, 5678.
399. Gopinathan, M. S.; Narasimhan, P. T. *Mol. Phys.* **1971**, *21*, 1141.
400. Cohen, H.; Benezra, C. *Org. Magn. Resonance* **1973**, *5*, 205.
401. White, D. W.; Verkade, J. G. *J. Magn. Resonance* **1970**, *3*, 111.
402. Hall, L. D.; Malcolm, R. B. *Can. J. Chem.* **1972**, *50*, 2092.
403. Hall, L. D.; Malcolm, R. B. *Chem. Ind. (London)* **1968**, 92.
404. Wasylishen, R.; Schaefer, T. *Can. J. Chem.* **1972**, *50*, 2989.
405. Terui, Y.; Aono, K.; Tori, K. *J. Am. Chem. Soc.* **1968**, *90*, 1069.
406. Schwarcz, J. A.; Perlin, A. S. *Can. J. Chem.* **1972**, *50*, 3667.
407. Wasylishen, R. S.; Schaefer, T. *Can. J. Chem.* **1972**, *50*, 2710.
408. Lemieux, R. U.; Nagabhushan, T. L.; Paul, B. *Can. J. Chem.* **1972**, *50*, 773.
409. Barfield, M.; Conn, S. A.; Marshall, J. L.; Miiller, D. E. *J. Am. Chem. Soc.* **1976**, *98*, 6253.
410. Doddrell, D.; Burfitt, I.; Grutzner, J. B.; Barfield, M. *J. Am. Chem. Soc.* **1974**, *96*, 1241.
411. Barfield, M. *J. Am. Chem. Soc.* **1980**, *102*, 1.
412. Smith, I. C. P.; Mantsch, H. H.; Lapper, R. D.; Deslauriers, R.; Schleich, T. "Conformation of Biological Molecules and Polymers," The Jerusalem Symposia on Quantum Chemistry and Biochemistry; D. Reidel Publishing Co.: Dodrecht, 1973, Vol. V, p 381.
413. Mantsch, H. H.; Smith, I. C. P.; *Biochem. Biophys. Res. Commun.* **1972**, *46*, 808.
414. Quin, L. D.; Gallagher, M. J.; Cunkle, C. T.; Chesnut, D. B. *J. Am. Chem. Soc.* **1980**, *102*, 3136.
415. Solkan, V. N.; Bystrov, V. F. *Izv. Akad. Nauk SSSR, Ser. Khim.* **1974**, 102; *Chem. Abstr.* **1974**, *80*, 121315s.
416. Solkan, V. N.; Bystrov, V. F. *Izv. Akad. Nauk SSSR, Ser. Khim.* **1974**, 1308; *Chem. Abstr.* **1974**, *81*, 136499z.
417. Ernst, R. R. *Mol. Phys.* **1969**, *16*, 241.
418. Hall, L. D.; Johnson, R. N.; Adamson, J.; Foster, A. B.; *Chem. Commun.* **1970**, 463.
419. Marshall, J. L. "Carbon–Carbon and Carbon–Proton NMR Couplings. Applications to

Organic Stereochemistry and Conformational Analysis"; Verlag Chemie International, Inc.: Deerfield Beach, Florida, in press, Chapter 3 and references cited therein.

420. Hansen, P. E. *Org. Magn. Resonance* **1978**, *11*, 215.
421. Bovey, F. A. "Nuclear Magnetic Resonance Spectroscopy"; Academic Press, Inc.: New York, 1969, Appendix E, Sections B and D.
422. Barrett, A. G. M.; Barton, D. H. R.; Nagubandi, S. *J. Chem. Soc., Perkin I* **1980**, 237.
423. Daniewski, W. M.; Griffin, C. E. *J. Org. Chem.* **1966**, *31*, 3236.
424. Ramey, K. C.; Moriarty, R. M.; Gopal, H.; Adams, T.; *Org. Magn. Resonance* **1969**, *1*, 101.
425. Flautt, T. J.; Erman, W. F. *J. Am. Chem. Soc.* **1963**, *85*, 3212.
426. Anet, F. A. L. *Can. J. Chem.* **1961**, *39*, 789.
427. Allen, C. F. H. *Can. J. Chem.* **1967**, *45*, 1201.
428. Shafi'ee, A.; Hite, G. *J. Org. Chem.* **1968**, *33*, 3435.
429. Sasaki, T.; Manabe, T.; Nishida, S. *J. Org. Chem.* **1980**, *45*, 479.
430. Sasaki, T.; Manabe, T.; Nishida, S. *J. Org. Chem.* **1980**, *45*, 476.
431. Mortimer, F. S. *J. Mol. Spectrosc.* **1959**, *3*, 528.
432. Günther, H.; Klose, H.; Cremer, D. *Chem. Ber.* **1971**, *104*, 3884.
433. Altona, C.; Sundaralingam, M.; *J. Am. Chem. Soc.* **1970**, *92*, 1995.
434. Kropp, P. J.; Krauss, H. J. *J. Am. Chem. Soc.* **1969**, *91*, 7466.
435. Booth, H. *Progr. NMR Spectrosc.* **1969**, *5*, 149, and references cited therein.
436. Haasnoot, C. A. G.; de Leeuw, F. A. A. M.; Altona, C. *Tetrahedron* **1980**, *36*, 2783.
437. Daniels, P. H.; Wong, J. L.; Atwood, J. L.; Canada, L. G.; Rogers, R. D. *J. Org. Chem.* **1980**, *45*, 435.
438. Barfield, M.; Marshall, J. L.; Canada, E. D., Jr. *J. Am. Chem. Soc* **1980**, *102*, 7.
439. Barfield, M.; Marshall, J. L.; Canada, E. D.; Willcott, M. R. III. *J. Am. Chem. Soc.* **1978**, *100*, 7075.
440. Wasylishen, R. E.; Barfield, M. *J. Am. Chem. Soc.* **1975**, *97*, 4545.
441. Govil, G. *Mol. Phys.* **1971**, *21*, 953.
442. Tori, K.; Iwata, T.; Aono, K.; Ohtsuru, M.; Nakagawa, T. *Chem. Pharm. Bull.* **1967**, *15*, 329.
443. Gassman, P. G.; Heckert, D. C. *J. Org. Chem.* **1965**, *30*, 2859.
444. Ogg, R. A.; Ray, J. D. *J. Chem. Phys.* **1957**, *26*, 1339, 1340.
445. Kawazoe, Y.; Tsuda, M.; Ohnishi, M. *Chem. Pharm. Bull.* **1967**, *15*, 214.
446. Märkl, G.; Lieb, F. *Angew. Chem., Internat. Edit. Engl.* **1968**, *7*, 733.
447. Hildenbrand, K.; Dreeskamp, H. *Z. Naturforsch.* **1973**, *28B*, 126.
448. Santini, C. C.; Fischer, J.; Mathey, F.; Mitschler, A. *J. Am. Chem. Soc.* **1980**, *102*, 5809.
449. Wray, V. *J. Am. Chem. Soc.* **1978**, *100*, 768.
450. Wray, V. *Progr. NMR Spectroscopy* **1979**, *13*, 177.
451. Marshall, J. L.; Faehl, L. G.; Kattner, R.; Hansen, P. E. *Org. Magn. Resonance* **1979**, *12*, 169.
452. Marshall, J. L.; Faehl, L. G.; Kattner, R. *Org. Magn. Resonance* **1979**, *12*, 163, and references cited therein.
453. Wray, V. *J. Chem. Soc., Perkin II* **1976**, 1598.
454. Williamson, K. L.; Braman, B. A. *J. Am. Chem. Soc.* **1967**, *89*, 6183.
455. Hirao, K.; Nakatsuji, H.; Kato, H.; Yonezawa, T. *J. Am. Chem. Soc.* **1972**, *94*, 4078.
456. Petrakis, L.; Sederholm, C. H. *J. Chem. Phys.* **1961**, *35*, 1243.
457. Petrakis, L.; Sederholm, C. H. *J. Chem. Phys.* **1962**, *36*, 1087.
458. Ng, S.; Sederholm, C. H. *J. Chem. Phys.* **1964**, *40*, 2090.
459. Hilton, J. H.; Sutcliffe, L. H. *Progr. NMR Spectrosc.* **1975**, *10*, 27.
460. Krespan, C. G. *J. Am. Chem. Soc.* **1961**, *83*, 3432.
461. Quin, L. D.; Mesch, K. A. *Org. Magn. Resonance* **1979**, *12*, 443.
462. Kitching, W.; Drew, G.; Adcock, W.; Abeywickrema, A. N. *J. Org. Chem.* **1981**, *46*, 2252.
463. Barfield, M.; Chakrabarti, B. *Chem. Rev.* **1969**, *69*, 757.
464. Sternhell, S. *Pure Appl. Chem.* **1964**, *14*, 15.
465. Jackman, L. M.; Sternhell, S. "Applications of NMR Spectroscopy in Organic Chemistry," 2nd ed.; Pergamon Press: New York, 1969, Chapter 4-4, pp. 334–341.
466. Barfield, M.; *J. Am. Chem. Soc.* **1975**, *97*, 1482.
467. Lehn, J. M.; Wipff, G. *Theor. Chim. Acta* **1973**, *28*, 223.
468. Lehn, J. M.; Wipff, G. *Tetrahedron Lett.* **1980**, *21*, 159.
469. Marchand, A. P.; Weimar, W. R., Jr. *J. Org. Chem.* **1969**, *34*, 1109.
470. Garbisch, E. W., Jr. *J. Am. Chem. Soc.* **1964**, *86*, 5561.
471. Tori, K.; Ohtsuru, M.; Hata, Y.; Tanida, H. *Chem. Commun.* **1968**, 1096.

472. Abraham, R. J.; Gottschalk, H.; Paulsen, H.; Thomas, W. A. *J. Chem. Soc.* **1965**, 6268.
473. Mark, V. *Tetrahedron Lett.* **1974**, 299.
474. Brouwer, H.; Stothers, J. B. *Can. J. Chem.* **1971**, *49*, 2152.
475. Meinwald, J.; Meinwald, Y. C. *J. Am. Chem. Soc.* **1963**, *85*, 2514.
476. Rassat, A.; Jefford, C. W.; Lehn, J. M.; Waegell, B. *Tetrahedron Lett.* **1964**, 233.
477. Tori, K.; Ohtsuru, M. *Chem. Commun.* **1966**, 886.
478. Baldwin, J. E.; Pinschmidt, R. K. *J. Am. Chem. Soc.* **1970**, *92*, 5247.
479. Adams, C. H. M.; Mackenzie, K. *J. Chem. Soc.* (*C*) **1969**, 480.
480. Cimarusti, C. M.; Wolinsky, J. *J. Org. Chem.* **1971**, *36*, 1871.
481. Garbisch, E. W., Jr. *Chem. Ind.* (*London*) **1964**, 1715.
482. Miller, R. G.; Stiles, M. *J. Am. Chem. Soc.*, **1963**, *85*, 1798.
483. Bystrov, V. F.; Stepanyants, A. U. *J. Mol. Spectrosc.* **1966**, *21*, 241.
484. Meinwald, J.; Lewis, A. *J. Am. Chem. Soc.* **1961**, *83*, 2769.
485. Padwa, A.; Sheffer, E.; Alexander, E. *J. Am. Chem. Soc.* **1968**, *90*, 3717.
486. Wiberg, K. B.; Lampman, G. M.; Siula, R. P.; Connor, D. S.; Schertler, P.; Lavanish, J. *Tetrahedron* **1965**, *21*, 2749.
487. Wiberg, K. B.; Connor, D. S. *J. Am. Chem. Soc.* **1966**, *88*, 4437.
488. Baker, K. M.; Davis, B. R. *Tetrahedron* **1968**, *24*, 1663.
489. Wasylishen, R.; Schaefer, T. *Can. J. Chem.* **1972**, *50*, 1852.
490. Peat, I. R.; Reynolds, W. F. *Can. J. Chem.* **1973**, *51*, 2968.
491. Anet, F. A. L.; Bourn, A. J. R.; Carter, P.; Winstein, S. *J. Am. Chem. Soc.* **1965**, *87*, 5249.
492. Coates, R. M.; Kirkpatrick, J. L. *J. Am. Chem. Soc.* **1968**, *90*, 4162.
493. Jefford, C. W.; Hill, D. T.; Ghosez, L.; Toppet, S.; Ramey, K. C. *J. Am. Chem. Soc.* **1969**, *91*, 1532.
494. Stock, L. M.; Wasielewski, M. R. *J. Org. Chem.* **1970**, *35*, 4240.
495. Gribble, G. W.; Kelly, W. J. *Tetrahedron Lett.* **1981**, *22*, 2475.
496. Benezra, C. *Tetrahedron Lett.* **1969**, 4471.

Author Index

---

* Each author's name is followed by the reference number. The numbers in parentheses represent the pages on which that reference number is cited. Complete literature citations appear on pages 207–217 in numerical order.

222 AUTHOR INDEX

Gil, V. M. S. **273** (65) (66) (67)
Gillies, D. G. **304** (75) (152)
Gladstone, C. M. **368** (91) (135)
Glick, R. E. **133** (31) (40), **137** (32)
Goering, H. L. **30** (9)
Gold, V. **3** (1)
Golden, R. **147** (32)
Goodin, D. B. **19** (4) (9)
Goodman, L. **138** (32)
Gopal, H. **356** (89) (113) (125) (126) (127) (183) (184), **424** (125)
Gopinathan, M. S. **399** (114)
Gorenstein, D. G. **113** (28) (75)
Gottschalk, H. **472** (191)
Govil, G. **441** (149)
Grant, D. M. **94** (23), **95** (23), **96** (23), **97** (23), **284** (71), **326** (96)
Green, M. **386** (106) (206)
Greene, R. L. **278** (66)
Gribble, G. W. **495** (198)
Griffin, C. E. **423** (118) (155)
Griffiths, L. **25** (9)
Grisdale, P. J. **153** (32)
Grivich, P. **4** (1) (14) (19) (29) (32)
Grover, S. H. **107** (26) (27), **108** (26) (27), **168** (37)
Grutzner, J. B. **80** (20) (23) (25) (26) (36) (41) (42) (66) (67) (68) (103) (162) (163) (202), **410** (114)
Gschwinder, W. **100** (23) (27)
Günther, H. **325** (113), **432** (140)
Gupta, B. D. **146** (32)
Gurudata, **135** (31) (33) (95) (96) (139) (140) (141) (142) (188) (189) (190)
Guthrie, J. P. **107** (26) (27)

Haasnoot, C. A. G. **436** (144) (145)
Hageman, H. J. **392** (114)
Hakli, H. **22** (8)
Hall, F. H. **376** (99) (114) (149) (151), **398** (114) (149) (151)
Hall, L. D. **380** (100) (114) (123) (155), **402** (114), **403** (114), **418** (115) (166) (168)
Hamer, G. K. **155** (33)
Hammett, L. P. **51** (14) (29), **52** (14) (29)
Hansch, C. **57** (14)
Hansen, P. E. **420** (113), **451** (162)
Hansen, R. S. **385** (106)
Harano, K. **367** (95)
Harden, R. C. **11** (4) (9)
Harrison, A. C. **75** (19)
Hart, F. A. **26** (9)
Hart, P. A. **41** (11)

Hartcher, R. **351** (93) (94) (97)
Haszeldine, R. N. **75** (19)
Hata, Y. **122** (30), **123** (30) (76) (80), **471** (191)
Havinga, E. **392** (114)
Hawkes, G. E. **23** (8), **24** (8)
Hayamizu, K. **330** (80), (117) (118)
Haywood-Farmer, J. **64** (16) (18) (19) (87)
Heckert, D. C. **443** (153)
Hehre, W. J. **115** (29)
Herring, F. G. **71** (19)
Hildenbrand, K. **283** (71), **447** (153)
Hill, D. T. **493** (197) (198)
Hilton, J. H. **459** (168)
Hinckley, C. C. **9** (4)
Hine, P. T. **277** (66) (67) (104) (163)
Hines, T. **125** (30) (31) (76) (80) (116) (175)
Hirao, K. **387** (109) (114) (115) (153) (166) (168) (199), **455** (166)
Hite, G. **428** (136) (137)
Hofer, O. **15** (4)
Hoffman, R. A. **48** (11)
Hogeveen, H. **195** (42) (48) (49) (62)
Holder, R. W. **16** (4)
Holubka, J. W. **176** (39)
Homer, J. **161** (35), **162** (35), **163** (35)
Hopla, R. E. **340** (82) (97) (113) (123)
Hossain, M. B. **19** (4) (9)
Hsu, Y.-F. L. **354** (81) (82) (121) (123) (176) (177), **355** (82) (121) (123) (177), **376** (99) (114) (149) (151)
Hügel, H. M. **242** (58)

Ihrig, A. M. **8** (1) (3), **377** (99) (114) (149) (150) (151)
Inagaki, F. **18** (4)
Inamota, N. **126** (30) (31) (32)
Inubushi, T. **318** (77) (78) (110) (172) (173), **319** (77) (79) (110) (169) (171) (172), **321** (78) (110) (111) (173) (174)
Ivanov, V. T. **393** (114)
Iwata, T. **442** (153)

Jackman, L. M. **465** (174)
Jacobus, N. C. **128** (31) (32) (81)
Jakobsen, H. J. **385** (106)
Janks, C. **27** (9)
Jautelat, M. **80** (20) (23) (25) (26) (36) (41) (42) (66) (67) (68) (103) (162) (163) (202)
Jefford, C. W. **76** (20) (21) (22) (28) (76), **476** (182), **493** (197) (198)
Jensen, W. A. **242** (58)
Jerina, D. M. **301** (73)

Subject Index